*All good things
are wild and free*

Walking by
Henry David Thoreau
(1862)

The Wildlife Year

Sally Coulthard

Illustrations by Gemma Koomen

How to Reconnect with
Nature in Every Season

quadrille

For Amelia

Introduction

As a child, I spent most of my time in our long, narrow garden.
Like so many suburban plots, it was beautifully tended near the
house and slowly slid towards wilderness at the back. There, among
the brambles and remains of bonfires, I'd spend hours digging holes,
making dens and collecting any evidence of animal life I could find.
Empty shells, dead moths, birds' feathers, even gruesome bits of
fur, teeth or bone. All of it was gold. Like a naturalist in training,
I'd classify them into groups, glue them onto sheets of paper and
label my finds with the earnestness of a museum curator.

Most people first encounter nature as children. Very young people
experience plants and animals with such a joyful lack of cynicism.
The stripes of a wriggling earthworm or the armour of a woodlouse,
a living fossil if ever there was one, are as thrilling as any lion or
tiger in a remote part of the world. Children also seem unusually
empathetic with wildlife. As adults, so many creatures become
'vermin' or 'weeds', pests to be exterminated. Little minds don't
think like that, presenting bunches of dandelions as proudly as a
florist's bouquet, or burying the remains of a dead mouse with all the
solemnity of a family funeral. It's glorious. And should be cherished.

I also think there's something to be learned from how children see
the world. Their endless curiosity and willingness to be in awe are
some of the most useful traits we can have as adults. Many of the
things we take for granted, such as the weather or the seasons, are
magical. We also experience plenty of nature without really knowing
anything about the science, biology or history behind it. Kids ask the
questions we have forgotten the answers to, or never even bothered
to learn in the first place. What is snow? Why does the moon change
shape? Where do birds go in winter? And why do leaves change
colour? For how much time is a tadpole a tadpole? How long do
bumblebees live?

Nature never ceases to amaze me. I often describe my writing as the intersection between nature, people and history – I'm fascinated by the relationship between us, our natural environment, and how that has changed over the centuries. In many ways, that relationship is more sophisticated than it has ever been – we now know so many things about how the natural world works. And yet, something has also been lost. Unlike our ancestors, few of us could, with any confidence, recognize the plants in a typical hedgerow or name more than a handful of birds that visit our gardens. Key moments that once defined the year, such as full moons, solstices or natural signs of seasonal change, have also lost their meaning.

This book is designed to be a gentle reintroduction to many of the seasonal events that mark the natural year. Whether I'm suggesting plants to look out for, ways to engage with wildlife, or simple explanations of remarkable natural events, my aim is to encourage everyone to rekindle their passion for nature. It's also a plea to think again about some of the animals and plants that people aren't naturally drawn towards. Anyone can love a honeybee, for example, but other pollinators such as hoverflies, moths and wasps are working equally hard in fields and orchards without any thanks.

Nature is right on your doorstep. Whether you're in the country, city or, like my childhood home, somewhere in between, there is wildlife everywhere. And while it's trickier to find in some places than others, there are some striking examples – such as peregrine falcons – of nature thriving in the most man-made of spaces. The more you look, the more you see. The more you see, the more you care. And, if you rediscover your child-like sense of wonder, so much the better.

Spring

The season of wild contrasts. Charles Dickens wrote in *Great Expectations*, 'It was one of those March days when the sun shines hot and the wind blows cold: when it is summer in the light, and winter in the shade.' He was right: perhaps no other time in the year is so unpredictable. It can blow, shine and downpour all in the same week, and still leave space for an afternoon of snow. And yet, for all its anarchy, spring is resolutely optimistic. Since time immemorial, our ancestors have welcomed its return. For the Celts, the start of spring was celebrated with Imbolc – a feast whose name may have been linked to the arrival of lambs in spring. The origins of the word are unclear – in the early Celtic tongues it may have meant 'in the belly' or, perhaps, 'ewe's milk' – a relic of our ancient pastoral heritage. The date for Imbolc was, at some point in more recent history, fixed at 1 February. But our ancestors probably took a more fluid approach to spring's arrival, looking instead for seasonal signs for reassurance. Blackthorn blossom, early butterflies and woodland flowers would have been unmistakable natural cues. So too the rekindling of the dawn chorus, bumblebee flights and tadpoles in transient pools.

In the modern world, spring is defined in a number of ways. For meteorologists, the year is a perfect cake, easily divided up into 12 slices. Three slices make up a season – in the northern hemisphere, spring is all of March, April and May. These three months sit together for a reason. Meteorologists have divided the seasons into groups of three months based on similarities in average monthly temperatures. While summer has the hottest three months, and winter the coldest, both spring and autumn sit beautifully, and temperately, in between these two extremes. For those who follow the stars, however, and prefer an astronomical explanation, spring begins on the vernal equinox – one of only two times of the year when the hours of daylight and darkness are equal. This usually happens around 20 March.

For centuries, however, people made up their own minds about whether spring had truly arrived or not. Cues from plants, animals and the weather were carefully watched to see whether the season had started. Chambers' *Book of Days*, a Victorian miscellany of seasonal folk lore, noted rather sweetly that, 'we can now plant our foot upon nine daisies, and not until that can be done do the old-fashioned country people believe that spring is really come. We have seen a grey-haired grandsire do this, and smile as he called to his old dame to count the daisies, and see that his foot fairly covered the proper number.' Other country dwellers looked to the appearance of the earthworm, a creature that tunnelled further underground to escape the depths of winter. Such was the significance of the earthworms' re-emergence, March's full moon was known as the 'worm moon'. Even the cuckoo's cry was hailed as a harbinger of spring.

But for most people, the true sign of spring was the welcome return of longer, warmer days. In Anglo-Saxon, spring was *lencten*, from *langa*, 'to lengthen', no doubt after the broadening daylight. For centuries, the word simply meant springtime, before the Church calendar borrowed 'Lent' and disguised its original meaning. Other medieval writers named the month after its cleansing, fresh start, calling it *prime temps* (first season) or *newe tyme*. But it was a name that first appeared in the fourteenth century that really caught everyday folk's imagination – *sprygyng tyme* perfectly fitted the season's explosion of buds bursting, lambs leaping and birds breaking into song. We still call it springtime today.

Mizzle

Language is both beautiful and elusive. At times, words are as clear as a spring dawn, their meaning crisp and precise. At others, they can be opaque or soft around the edges, their definition ever shifting. 'Mizzle' is one of the latter. In use since at least the seventeenth century, its meaning is as vaporous as the weather it describes. For some it captures that drenching slow rain we often get at this time of year, somewhere between fog and drizzle. In some parts of the country, it's a thick, saturating pea-souper.

It sounds like a made-up word – a modern marriage of mist and drizzle – but its origins are probably older, and much odder. Mizzle may come from an ancient root word *meigh*, meaning to wet through or urinate: an etymological nugget that also turns up much later in other words such as mist, mistletoe and even micturate. Rather wonderfully, mizzle was also used as a verb – 'to mizzle' – and meant to run away, as if disappearing into the ether. Sneaky people, fantastically clever at effecting escapes or getting out of tight spots, were even known as rum-mizzlers, a seasonally evocative slur if there ever was one.

Birds' nests

Birds are the artisans of the natural world. Their nests are creations of extraordinary skill and inventiveness, transforming the flotsam of the forest floor and city park into homes of breathtaking beauty. Nests are also vernacular creations, objects crafted from local materials. From leaves and cobwebs to reeds and horsehair, birds' nests give a glimpse of the local ecosystem and what's available to pilfer.

Blueprints vary between species. Robins, song thrushes and blackbirds follow similar plans and build neat, tiny bowls of woven grass, moss and thin twigs in under a fortnight. Others, such as swallows and swifts, use the wattle-and-daub method, mixing soft mud or saliva with organic material that hardens in the sun. Long-tailed tits build the snuggliest nests of all. Oval in shape, like a mossy Easter egg, their homes are carefully woven from spiders' cobwebs, lichen and hair. Up to 2,000 tiny feathers are then used to plump the interior, a warm, soft nursery for delicate eggs and new hatchlings. Such craftsmanship takes time – long-tailed tits can spend a month completing their architectural masterpieces.

For the nature lover, however, nests are forbidden treasure. Decades of wild egg collecting by moustached Victorians decimated bird numbers. Rows of jewelled, speckled eggs were amassed in dusty cabinets, never to hatch into baby birds. Thankfully, society has seen sense and nests and their contents are now protected by law. In fact, wild birds need all the help we can give them. Many species are declining, partly because birds are increasingly struggling for nesting materials and places to build a home. We can, however, help.

How you garden affects a wild bird's ability to nest. Many birds love scrub, that deliciously chaotic blend of brambles, dog rose and elder that often thrives on patches of unkempt land. Muddy areas and pond margins are vital for swifts and swallows. Don't clear away all the leaf litter, moss and small twigs – they're a bird's tools of the trade. Provide hanging baskets or metal suet cages of nest-building materials during spring – feathers, wool, moss, coconut fibre, and short pieces of grass or straw of varying lengths.

A rich array of hedges, bushes and mature trees is also key, and so is their maintenance. Hedge trimming and pruning should wait until autumn, when birds have finished raising their brood. Plant a mix of native trees, evergreens and dense, deciduous hedgerow if you have space. Nest boxes can also provide suitable habitats for birds when natural nesting sites are scarce. Different species need different designs – robins prefer open-fronted boxes just 1–2m (3–7ft) off the ground, for example, while swifts prefer the high life, nesting in the crevices and eaves of houses and outbuildings. Make your house a welcome home for birds – new builds and renovations can even incorporate permanent nest boxes, such as swift bricks, sparrow blocks or owl holes into their very fabric. Come spring and you'll have avian house guests flocking.

Hares

Early spring is the breeding season for hares and the perfect time of year to spot two pugilists at work. For centuries, people assumed that boxing – or 'mad March' – hares were two males, fighting to win the affection of a winsome female. In truth, it's the female who instigates a boxing match, raising up on her hind legs and raining down blows on a potential suitor. Why female hares start a fight isn't fully understood. One theory is that the female is fending off the unwanted attentions of an over-zealous male, who won't take no for an answer, while others suspect she could be testing his mettle prior to mating. Either way, for a ring-side seat, head to an open stretch of arable land at dawn or dusk during March. Bring binoculars and stay downwind – hares have sublime hearing and an even better sense of smell.

Rain

Why do we say 'April showers'? Although April isn't usually the wettest month of the year (October to December is traditionally the soggiest stretch), April is particularly unsettled when it comes to the weather. The position of the jet stream, a strong current of air that blows high above the Earth's surface, influences the meteorological conditions we have. In April, it moves northwards, bringing with it squally rain and wind from over the Atlantic.

April is also the time of year when the sun begins to rewarm the northern hemisphere. When water in rivers, streams and oceans is gently heated it turns to water vapour. This moist warm air rises skywards, gradually cooling and condensing as it gets higher. Moisture in the air will find tiny dust particles, called condensation nuclei, and cling to them. If there are enough of these small water drops together, they become visible as clouds. As these small water droplets grow and coalesce to form bigger drops, they become heavy and fall to the ground. This is rain.

Remarkably, the shape of a raindrop is not what you'd imagine. Far from a teardrop, a raindrop usually takes the shape of a hamburger bun. A raindrop, just before it falls from a cloud, is spherical. As it plummets towards the ground, wind resistance flattens the bottom of the raindrop, creating a water droplet with a flat base and domed top. Larger raindrops can even start to form a different shape as they reach terminal velocity. The hamburger starts to narrow in the middle, creating a jellybean shape. If a jellybean raindrop gets really big, it can split into two new raindrops. The only reason we expect raindrops to be tear-shaped is because most water droplets we see are drips. When a droplet forms at the end of your bathroom tap, for example, it keeps getting bigger and stretching downwards, due to gravity. At some point the surface tension, which is keeping the droplet attached to the tap, becomes too weak and the droplet, well, drips.

Frogs & frogspawn

Humans love to sentimentalize spring, but nature knows its true purpose. The race to mate is an existential one and perhaps none feels it so acutely as the common frog. Longer daylight hours and the increasingly clement weather tell the frog that it's time to go back to the pond of their birth and breed. For some it's just a short hop across a garden or field; for others a hike of half a kilometre (a third of a mile). As they arrive in their dozens, males scramble to find a mate among the mud and weeds. The lucky few will grip their bride tightly, in a piggyback hold called 'amplexus', and fertilize her eggs as she releases them into the water.

From that point onwards, each embryo is self-sufficient, fighting for survival alongside its brothers and sisters. Unlike the common toad, which lays its spawn in elegant ribbons draped among the pond weeds, the common frog leaves a glutinous blancmange. Over the next fortnight, the tiny embryos quickly grow – from a black full stop to a wriggling comma – before bursting out of their jelly capsules and swimming free. The pond remains the tadpoles' nursery for the next three months but there's no time for play. Each must undergo one of nature's most extraordinary transformations – a complete metamorphosis from an underwater creature to a land-dwelling frog. Gills become lungs, soft tissue hardens into bone, and intestines, jaw and mouth slowly shift from an algae-based diet to a meat-eating menu. First back- and then front-legs grow, ready for terrestrial life. Even the long tail, the wriggling tadpole's pride and joy, is reabsorbed in the final few days before the new baby froglet hops off into the damp undergrowth.

The common frog's Latin name, *Rana temporaria* or 'temporary frog', perfectly sums up this shy creature's habits. After the frenzy of spring breeding and tadpoles, it sneaks off and seems to disappear. The common frog also tends to hunt at night or late in the day, snatching slugs, flies and earthworms with its long, sticky tongue, staying well away from the scorching sun. While most frogs only use ponds for the spring breeding season, they'll spend the rest of the year relatively close by, foraging in damp, earthy places such as leaf litter, ditches, under hedges and logs, rock piles and compost heaps. A garden filled with dark, moist corners and a pond, however small, will be perfect for frogs. Amphibians are also particularly susceptible to toxic chemicals, which are easily absorbed through their highly permeable skin. Avoid using weedkillers, pesticides and slug pellets if you want an anuran-friendly patch. With numbers of common frogs declining across Britain and Europe, due in no small part to habitat loss and lack of breeding ponds, there's never been a better time to make a frog feel at home.

Toads

The common toad is a wanderer at heart. Come springtime, toads make long return journeys to their ancestral pools to breed, often crawling as far as 5km (3 miles) in search of romance. It's a perilous voyage, however, not least because large numbers of toads must cross busy roads, at specific places, to reach their watery destinations. Traffic deaths account for a large proportion of toad fatalities but everyone can help. 'Toad patrols' are groups of volunteers who monitor well-used stretches of road during spring evenings, collecting amphibians from one side and ferrying them across in buckets. Patrols also monitor toad numbers, and places where toads traditionally congregate. This data can be fed back to local government and conservation bodies so road signs can be erected or, even better, amphibian tunnels built to allow toads and other animals to continue their roving in safety.

Kingfishers

Few sights are as dazzling as the lucent turquoise and tangerine flash of a kingfisher. These tiny, robin-sized birds lay their first clutch of eggs in March or April in a burrow tunnelled along the water's edge. It takes around three weeks for the chicks to hatch and, from then on, it's a race for both parents to stuff their youngsters to the brim. Spring is the perfect time to sit patiently at a site with slow-moving water, hoping to catch a flicker of this regal bird darting for small fish, tadpoles and dragonfly nymphs. Head to your nearest river, stream, lake, canal or wetland – kingfishers love to perch on overhanging branches, returning to the same spot again and again. Kingfishers are out and about throughout the day but early mornings or after a heavy rainfall are optimum times, when busy parents are dashing to fill young, hungry stomachs.

Moles & molehills

We know very little about the velvety mole. This pocket-sized, secretive mammal spends most of its time below ground, out of the limelight and away from prying eyes. But, during spring, we get a glimpse of this creature's assiduous burrowing in the form of molehills pockmarking the lawn.

Molehills are only created when new tunnels are dug. For most of the year the mole keeps itself to itself, squeezing its way through an established underground network. In springtime, however, male moles extend their tunnels in search of subterranean romance, hoping to find a female in the crumbly darkness. Moles also make new burrows as their favourite food heads nearer the surface. In winter, earthworms escape the freezing temperatures by burrowing further down into the soil. Come the gentle warmth of spring, they migrate upwards once again, with moles hot on their heels.

In fact, moles have such a singular appetite for earthworms that much of their life is dedicated to hunting them. Moles catch their prey by constantly patrolling tunnels, waiting to feel the vibration of an approaching earthworm. During spring, moles have their pick of juicy wrigglers but, when the summer drought hits, earthworms can go into a period of dormancy, making them harder to locate. Faced with a potential food shortage, during springtime moles catch more earthworms than they need and store them in 'living larders'. A toxin in the mole's saliva can paralyze an earthworm with one bite but, crucially, doesn't kill it outright. Stored underground in a cache with its colleagues, the hapless, immobilized earthworm will stay fresh for weeks, ready to be eaten when the mole returns.

Molehills might be a common sign of spring but, for centuries, people have raged against these miniature soil heaps and the creatures that construct them. Few realize, however, that molehill soil is gold-dust for the garden. With its high organic content, and loose texture, it makes sublime potting compost and top dressing. Moles are also clandestine heroes, gobbling up garden pests including wireworms, leatherjackets, chafer grubs and carrot-fly larvae, without asking for any reward in return.

Blossom

After the sombre quiescence of winter, spring's carnival arrives with an explosion of blossom. Resolutely cheerful, it's perhaps no surprise the word is both a noun and a verb. Blossom is not only a botanical term, used for the flowers of stone-fruit trees such as apples and cherries, but is also an action. Few descriptions conjure up such rich promise as 'to blossom'; to bloom and thrive like a spring floral display.

For centuries, people have heralded the return of blossom as the first sign of spring, a reassurance that bountiful times are ahead. Like an orchestrated dance, fruiting trees and hedgerows bloom in sequence, filling the season with their effervescent beauty. First to show their fineries are the cherry plum and blackthorn, both of which can burst into bleached blossom as early as February. Next in the wedding entourage are pear and plum, March's best-dressed fruit trees, also in perfect whites. April brings the crab apple to the party, along with bird cherry and wild cherry. Come May, and the blossoms are flushed with pink: the hawthorn, sour cherry and apple.

But blossom is not only nature's window dressing. The fortunes of native fruit trees and flying insects are inextricably linked. To produce plenty of fruit, a blossom tree's pollen must be couriered from one flower to the next. While honeybees often get the plaudits for pollination, many other species work just as hard away from the limelight. Bumblebees, solitary bees, wasps, and plenty other winged insects make huge contributions. Hawthorn hedgerows, for example, are kept buzzing mostly by flies, hoverflies and tiny beetles. Bumblebees, bristly flies and solitary bees play the leading

role in the pollination of blackthorn blossom. Even in commercial apple orchards, wild pollinators outperform honeybees. The often-overlooked hoverfly, for example, works at double-speed, visiting more blossom flowers than any other insect, closely followed by the bumblebee. Equally, what they lack in pace, solitary bees more than make up for in heft, carrying significantly more pollen at any one time than their hive cousins.

In return, spring blossom provides pollinating insects with both nectar and pollen. Many insects rely on fruit trees and hedgerows for one or both of these valuable foods. While nectar – nature's very own energy drink – provides a sugar-rich boost, pollen is packed with protein and other nutrients. Many bee species, including honeybees, also bring pollen back to the nest or hive, where it's fed to developing bees or stored for future use.

Insects and tree species coevolved. Different periods of blossom coincide with specific groups of pollinators, a mutually beneficial relationship that has been refined over millions of years. Bees that emerge very early in the season for example, such as the ashy mining bee and the white-tailed bumblebee, make a beeline for blossom, often one of the few food sources available before flowers have had chance to bloom. Gardens, parks and other communal spaces can help a wide variety of pollinating insects, and the birds that rely on them to feed their chicks. Planting a healthy mix of native stone-fruit species, ones that peak at different moments throughout spring and early summer, will ensure wildlife blossoms too.

Queen bumblebees

All hail the queen bumblebee. Like sleeping beauty, she has spent the entire winter hibernating alone. The kiss of spring warmth awakens her, and she'll now leave her home – usually a small hole in the soil – and set off to cram herself full with nectar and pollen. Once sated, the queen seeks out a nesting site – an old mouse hole perhaps or an abandoned bird box – and builds a tiny golden pot from wax, which she fills with nectar. She also constructs a small pea-sized mound of pollen and wax, onto which she carefully lays her first brood of eggs.

The queen's golden nectar pot provides all the sustenance she'll need for four days of brooding, sitting on her eggs waiting for them to turn into larvae. Once hatched, the larvae need feeding too – a rich diet of pollen and nectar – all brought in by the lone queen and her endless sorties. A fortnight of feeding and the larvae will transform into cocoons. Another two weeks later and the queen's cossetted offspring – a small staff of female bumblebees – emerge, ready to be put to work.

From this point onwards, the queen bumblebee won't leave her nest again. Instead, she'll sit tight and boss her female workers around, who tirelessly clean the nest and forage for food. Later in summer, the queen will lay eggs again – this time producing only male bees and new queens – all of whom will leave their dying mother before the year is out. While the males' only job is to mate and die before winter, ensuring the survival of the colony, the new queens will hibernate and live to see the following year. Come the warmth of spring, they'll wake up and emerge, beginning the queen bumblebee's 12-month life cycle once again. The queen is dead. Long live the queen.

Dawn chorus

The dawn chorus must be one of the most democratic of all nature's encounters. Almost every corner of the country thrums with early-morning birdsong come springtime. All you have to do is open your window. Even before the sun has had a chance to light up the morning sky, many different species of bird have launched into full-throated melodies. From the chiffchaff's two-note repetition to the fluty ad-libbing of the song thrush, April and May are the best months to experience the sheer breadth and flair of avian chirruping. Although not a precise sequence, each species seems to have a different time slot on the morning's billing. First up to sing are the robin and blackbird, closely followed by thrushes, wrens, warblers and wood pigeons. Tits, sparrows, finches and other late-risers only chime in once it's light and many will continue to sing well after breakfast has been cleared away.

For all its musical charm, however, the dawn chorus is more of a hustings than a concert. Spring's lengthening days trigger a rush of hormones that tell songbirds it's time to find a mate. Males strike up early morning refrains as part of this reproductive game plan. Not only do they croon to defend their territories and warn off potential rivals, it's also a chance to impress a female with their musical virtuosity. The subtle meanings of the dawn chorus are still not fully understood – while some songbirds stop performing after successfully pairing up, others continue to shout from the rooftops or change their tunes, even with a nestful of new chicks back home.

Remarkably, although many birds' calls are innate, it seems that the melodies of the dawn chorus can only be learned. Young birds must be tutored in the art of singing and composition. Juveniles have to listen carefully to the songs of their elders, memorize the notes and patterns, and then go off and practice. Like eager undergraduates, the young songbirds then use this information to shape their own unique ballads as they grow up.

Ornithologists also suspect that, although bird songs within species sound identical to human ears, there is probably a huge amount of fine tuning and personalization in each individual's song. Every bird does its own cover version.

While twitchers might be able to easily discern the voices of different songbirds, the dawn chorus can seem like a cacophony to the untutored ear. Four birds, however, are easiest to pick out from among the choir. The house sparrow is brash – less of a song, more of an abrupt series of loud cheeps lasting for minutes at a time. The blackbird, on the other hand, does a wonderful jazz improvisation of three-second scats on a piccolo. The great tit chirrups a simple see-saw 'tea-cha tea-cha-tea-cha tea-cha', followed by a brief pause before repeating it again. And the wood pigeon, although not strictly a songbird, often cuts through the noise with its soothing 'coo-waaa-waaa wa wa'.

Swifts

For centuries, one bird went by many names. Its behaviour was deeply mystical, seeming to arrive from nowhere in large numbers only to vanish just a few weeks later. With its death-defying wheeling and breakneck speed, it commanded the sky without ever touching the ground. Even its piercing whistles and dense, dark plumage invited comment. For our ancestors, this bird was known as the swing devil, black martin, or satan's screecher. Those who thought it had no legs and toes called it the martlet, after the mythical bird that could never land. Even its scientific name, *Apodidae*, meant 'without feet'. For others, however, only one name suited this ravishingly quick bird: swift.

Come late April and early May, the swift returns. It makes a journey of unknowable heroism, flying distances of between 8,000 and 14,000km (5,000 and 9,000 miles) from its home in sub-Saharan Africa. Far from taking a leisurely pace, the swift lives up to its name and rips through 800km (500 miles) a day, eating, sleeping and even mating on the wing. It flocks north to breed, stopping at favourite places and often reusing the same nest year after year. There, it'll raise a small family, before heading back south later in the summer.

How devastating, therefore, to find that when they arrive, many swifts' homes have disappeared. The rise in house renovations, barn conversions and demolitions of old buildings means the places where swifts traditionally nest are being blocked up. Swifts love a nook and cranny, a small hole leading to a larger void – under a roof tile, for example, or inside a wall cavity. To prevent a further drop in swift numbers – which have more than halved in just two decades – we need to be hospitable hosts. Make or buy a swift box. Even better, install a permanent swift brick into your wall – town and city homes can offer just as much as the countryside for our frequent flyers. Swifts are also affected by a decline of insect prey – ensure your outside space offers a smorgasbord of flying bugs by planting generously for pollinators, creating a pond, leaving areas wild and ditching the chemicals. And, do it swiftly.

Slugs

It's time to call a truce on the slug. While most gardeners regard all slugs as a menace, only a few species do any damage. In Britain, for example, of the 44 recognized slug species, just nine munch garden plants. The rest eat rotting vegetation, algae, food waste, and dead flesh. Some slugs even eat other slugs. The slender long worm slug, for example, usually sticks to carrion, earthworm poo and decaying plants. Equally, the speckly green cellar slug shuns petunias and prefers a diet of mould and algae. It's also partial to pet food and the slug you'll mostly likely find indoors, heading for the dog bowl. The rather pretty, ear shelled slug, so named after the tiny fingernail-like shell on the tip of its tail, survives exclusively on earthworms and would never dream of attacking your lettuces. And the glorious leopard slug, a whopper with unmistakable big-cat markings, stalks other slugs at night, chasing and taking them down at a giddy pace of 15cm (6in) a minute.

Slugs are also a vital part of other creatures' diets. Different species of garden birds, frogs, toads, hedgehogs, badgers, foxes, slow worms, and beetles eat slugs, some more enthusiastically than others. Hedgehogs tend to turn to slugs as a last resort, often when other food sources are scarce, while both frogs and toads gobble down as many as they can find. For some creatures, such as the little owl and tawny owl, slugs form part of a wide and varied diet. Others, such as slow worms, heavily rely on the gastropod for survival, partly because it's one of the few animals that's unhurried enough to catch. Even the slug's natural defence system – its foul-tasting, sticky mucus – isn't enough to deter its epicurean fans. Song thrushes often wipe a slug to and fro on the ground, sloughing off the unpalatable slime before guzzling it down or taking it back to the nest. Hedgehogs, too, have been observed both rolling slugs in dirt or wiping them with both front paws before tucking in.

Rabbits

Rabbits are ancient symbols of new life and fertility, the perfect emblem for spring babies and Easter. In reality, rabbits produce litters throughout the year, only pausing during deepest winter. A female rabbit is usually ready for motherhood at only three months old. Short pregnancies of only a month, and litters of up to 12 new kits at any one time, mean a doe can have as many as 60 babies a year. Without even pausing for maternity leave, she can also become pregnant again only one day after giving birth. It's a deeply sensible parenting strategy, however, in the face of high mortality rates. Over 90 per cent of young rabbits die in their first year of life, mostly in the first three months. Those that come through unscathed can live to three years old but rarely more. Thankfully, there'll always be more kits just around the corner.

Blooms & bees

The relationship between bees and flowers is a match made in heaven. And one of the most extraordinary markers of this long-lasting love affair is a behaviour known as 'buzz pollination'. Some flowers are particular about who they want to pollinate them. Only certain species of bee will do – those that the flower knows will do a first-rate job of collecting and transferring the limited amount of pollen it produces. If a fly or wasp, for example, lands on a buzz-pollinated flower, they won't be able to access the pollen. When a bee lands, however, it performs an amazing trick. The bee grabs the anther in the centre of the flower – the part of the stamen that holds the pollen – and begins to vibrate wildly. It does this by using the strong muscles that usually power its wings. The high-frequency vibrations cause the flower to release its pollen in a powdery cloud. The dusted bee then combs her body, pushing the pollen onto her back legs before flying away. And, while bumblebees and certain species of solitary bee have buzz pollination down to a fine art, honeybees can't do it. Quite why, no one knows.

In fact, the relationship between flowering plants and all kinds of bee is truly fascinating and one we're only just beginning to comprehend. Certain bees, for example, can only make use of certain plants. Some bees have really short tongues, for instance, and can only drink nectar from shallow flowers such as daisies. Other have long tongues, designed for slurping the nectar from long, tubular blooms such as snapdragons. Some short-tongued species, when thwarted by a flower that's too deep, cheat and nibble a hole, sucking out the nectar from the side.

Even less well understood is how plants attract certain insects. Around half of all flowering plants, for example, are thought to lace their nectar with caffeine. This sneaky drink-spiking isn't designed to give its imbiber an energy boost, but rather to sharpen the bee's ability to remember the flower's distinctive scent. Other varieties of flower secrete nicotine into their nectar, an addictive substance that speeds up a bee's ability to learn flower colours. Flowers also have electrical fields that differ depending on the shape of the bloom, helping bees distinguish between flower varieties. The electrical field of a flower also changes briefly after it has been visited by a bee, before returning back to normal. No one knows for sure but this may be a way for the flower to signal that it's briefly short of nectar and just needs a little time to make more.

What is certain is that flower diversity is critical to keeping bee populations healthy. And, while we traditionally associate rural areas with lots of flowers, modern farming methods mean urban gardens often offer a much greater variety of blooms. Spring is the perfect time to introduce a rich mix of flowering plants, ones that will come into bloom at different times and feature a variety of flower heads, perfect for every type of bee.

Stag beetle

Few creatures love a spring clean as much as beetles. Busy housekeepers, many species are responsible for clearing up nature's detritus such as decaying plants, fungi and creature corpses. Others greedily devour garden pests, such as aphids or mildew. One of the most impressive is the stag beetle. Named after its distinctive 'antlers', the stag beetle's projections are actually its jaws and are used not for eating or biting, but for locking horns with other rival males. In fact, the male's antlers are so cumbersome that they not only prevent it from eating anything but tree sap, but also threaten to topple it over if it tries to run. Now endangered in many areas, the stag beetle relies on rotting timber. Its larvae need decomposing logs and old broadleaved trees to live in and feed on. Protecting ancient woodland, planting native trees, and creating piles of decaying wood all help safeguard this charismatic species.

Summer

In many historic cultures, the four seasons of the year mirrored the four stages of life. While spring was birth, autumn adulthood, and winter senescence, summer was the year's adolescence. It's an excellent metaphor. Summer is the season of bountiful energy and marked growth. It's also wonderfully and deliciously out of control. Just like a teenager, summer rarely does what it's told and careers between blistering sun and unpredictable downpours. Plants grow like weeds, animal hormones rage, and the days stretch, blissfully, into warm, dozy evenings.

The three months of summer – June, July and August in the northern hemisphere – also go through their own unique progression. June, with its chill nights, often has more in common with its spring predecessors but soon gets into its stride. Its greatest event, however, is the summer solstice, the longest day of the year. Since our earliest ancestors first picked up the plough and planted crops, midsummer was a critical time in the year. Neolithic monuments such as Stonehenge, which align with the movements of the sun, gave our ancestors a focal point for rituals and celebrations to mark this key event in the solar calendar.

June was named after Juno, the Roman goddess of women and childbirth. Her name means 'youthful vigour', a perfect etymology for such a lively month. July follows with equal bounding enthusiasm – the Romans again gave it its modern name, from Julius Caesar. But the Celts had a better phrase – *mìos buidhe*, meaning yellow month, a glorious recognition of July's golden fields of corn, hay and generous sunshine.

And then August, the month where nature's potential peaks and farmers begin their harvest. Anglo-Saxons called it *woedmonath*, or weed month, no doubt in recognition of nature running amok. First August was also Lammas Day, or 'loaf-mass' day, a festival that marked the beginning of the harvest with warm bread and bonfires. August had its dangers too. St Roch's feast day falls in the middle of the month. During the fourteenth century Roch, a pilgrim, put himself in grave danger to tend to victims of the plague. His kind works earned him a sainthood and his feast day sits at the perfect time of the year, August, the most likely month for pestilence and epidemics. In these hot weeks, many herbalists took advantage of the month's profusion of 'weeds' to create healing drinks – sage, yarrow, tansy and, of course, feverfew.

Of all the August weeds, however, perhaps none is more loved than the common poppy. Also called the corn poppy, it has long been a bedfellow of agricultural crops. Poppies learned to thrive where farmers disturbed the soil, timing their flowering and seed fall just before the late summer crops were scythed and stooked. They also resolutely refused to be picked and owned; poppies famously flop almost as soon as you cut them. Bringing a bunch of poppies indoors was also considered bad luck, an invitation for thunder and lightning to spoil the summer's bountiful harvest.

Gloaming

The softly lit minutes between darkness and daylight are known as twilight. For centuries, people have been mesmerized by this strange, nebulous time of the day. Neither fully light, nor fully black, twilight is the last of the sun's rays refracted by our dusty atmosphere, leaving the sky lit even when the sun has dipped below the horizon.

Summer is the season for half-light. Twilight extends on these long balmy days and its peace and beauty have long inspired lovelorn poets and artists. As one anonymous Victorian penned, 'It is the hour when young lovers wander through the green lanes between the hawthorn and the clematis, the nightingale sings high in the elm-tree, and the white moths flit like winged ghosts [...] Everything is dreamy, indeterminate, and full of possibilities not yet realized.'

Different cultures have given twilight its own redolent name. For the French, it's *l'heure bleu* (the blue hour) or *l'heure du berger* (the shepherds' hour), when hill sheep were at their most vulnerable to predators. In Latin, twilight was *inter canem et lupum* (between dog and wolf). But 'gloaming' – a word still used in Yorkshire and Scotland – outshines them all, a vestige from the Old English *glowan*, the light and lustre of red-hot metal.

The summer sky

The summer sky is an Impressionist's palette. By mid-morning, the heavens are a wash of cerulean blue. Come the day's end, the sky darkens from the top down, turning dusk almost ultramarine before slipping into blackness. In between, our crepuscular skies fuse one shade into another – twilight blending crimsons into tangerines and pale lemons into salmon pinks. Even the rain can't stop the sky's creative flourishes – summer rainbows appear in broad brush strokes – always fleeting and frustratingly out of reach.

Our ancestors marvelled at the summer sky's kaleidoscope, inventing gods and goblins to explain its beauty. But the real science is even more engrossing. Although it appears white to our eyes, sunlight is made of all the colours of the rainbow. This light travels in waves, with each colour having a different wavelength. Red has the longest wavelength, while blue is shorter. When sunlight reaches the Earth's atmosphere, the waves of coloured light begin to be scattered by tiny gas molecules in the air. While most of the longer, redder wavelengths can pass straight through these gas molecules, the shorter, bluer wavelengths are sent crashing in all directions, making the sky appear azure.

The angle of sunlight as it enters the atmosphere affects the colour of the sky. At sunset and sunrise, light must travel through more of the atmosphere than it does during the day. This longer journey results in more of the light being scattered, including longer wavelengths, making the sky appear red and orange. Summer evenings are also prone to rouged skies because the air contains more tiny dust and pollen particles, which also scatter long-wave light.

Rainbows are white light scattered into its constituent colours. As sunlight hits rain or mist, the spherical droplets act as prisms, separating the sunlight into its different hues. The rainbow doesn't exist without the viewer, however. To see one, the source of water needs to be in front and the sun must be shining from behind your back. It also needs to be low in the sky. Only then will the rainbow reveal itself in all its technicolour glory.

Fledglings

It's a tricky business, learning to fly. Come early summer and you may find baby birds, or fledglings, flitting and tumbling around your lawn. And, although they appear to be grounded, fledglings are usually exactly where they should be. Many garden birds leave the nest a few days before they can fully fly. Often looking slightly dishevelled, and fluffy around the edges, fledglings spend a few days flight training at ground level, perfecting their technique and growing their final flight feathers, before full take-off can commence. Far from being abandoned, the fledging's parents are probably watching nervously from nearby. Unless in imminent danger, leave fledglings where they are. If you find a fledgling on a busy road, in danger of predation or in an exposed location, however, gently carry the tiny flyer a short distance to safety. Also be sure to keep any pets indoors.

Adders

Few sights are as arresting as the glimpse of an adder, one of only
a handful of venomous snakes in Europe and the only one in Britain.
Cold-blooded, they long for the heat of summer and at this time of
year spend hours basking in its gentle glow. Although reclusive for
much of the year, they're often found lounging around woodland
rides, dunes, heath and moorland during this season, soaking up
the sun's rays. The males and females are unmistakable. Geometric,
like a Victorian tiled hallway, the boys dazzle with their silver-black
zig zags, while the girls opt for tan and dark brown. You may even
find them in a tight embrace, tangled together like coiled rope. Just
don't get too close; while timid by nature, adders will lash out in self-
defence if handled or trodden on. If you see one, admire it from afar
and let it slither off into the undergrowth.

Verges

One of nature's last wild refuges can be found in a surprising place. At the side of many roads, verges are scraps of grassland and hedgerow that offer a boundary between traffic and field. In summer, they foam with wildflowers, grasses and rambling climbers, and are often filled with native plant species – such as harebells and field scabious – that struggle in more manicured parts of the country. Changes in land use, and agricultural intensification, have left road verges as one of the last bastions for wildlife. As animals' habitats become more fragmented, verges remain important corridors through the landscape, safe stretches of land where they can forage for food, find a mate and connect to other territories.

When totted up, the amount of land dedicated to road verges is significant. In Britain alone, there are over 480,000km (300,000 miles) of rural road verges that, if managed properly, could be home to billions of wildflowers, insects, birds and bats. Verges also differ in their plant composition across the landscape – woodland verges bloom with shade-loving foxgloves, bluebells and wood anemone, while heathland verges provide flashes of heather, bilberry and heath bedstraw. Lowland verges are alive with vetches, wild orchids and oxeye daisies. And urban verges, often a welcome relief in a built-up landscape, can thrum with pollinator-friendly dandelions, daisies and yarrow.

The key to healthy verges is management. While each type of verge needs slightly different treatment, in general, cutting too early in the season, before most of the wildflowers have had chance to seed, reduces the amount of species that will thrive. For most areas, late summer is probably the optimum time. Not cutting verges at all is even worse, as single plant species will tend to dominate and floral diversity will take a nosedive. Other strategies, such as removing grass clippings to stop the soil becoming too nutrient-rich, or only cutting the front of a verge but leaving the back to be trimmed biannually, encourages a wider range of plant heights, wildflower species and insects.

Local authorities, highway departments or parish councils sometimes need a gentle shove when it comes to managing road verges for wildlife properly. Creating a volunteer group, getting community support, and persuading your council to cut later and less often, is not only cost-effective but boosts biodiversity. While some people love the aesthetics of tidy, close-shaved verges, our mind-set needs to shift. Short-mown grass provides little in terms of wildlife or plant biodiversity. Road verges could be one of our most successful, and cost-saving schemes for rewilding spaces. If we want nature to thrive, we need to let it grow.

Earthworms

If summer temperatures get too blistering, and the soil begins to
dry out, the common earthworm must take drastic action. It goes
into a short period of 'sleep' called aestivation. The word comes from
the Latin *aestivatus* – 'to spend the summer' – a cheery etymology
for a rather stressful period for a wriggly worm. Without moisture,
an earthworm soon crisps up and dies. It breathes through its slimy
casing, a feat only possible because of the blood flowing close
to its surface and the skin remaining damp. To keep itself from
desiccating, the earthworm coils into a tight knot and seals itself
into a small, tight chamber deep underground. This self-enforced
summer isolation cell, lined with the earthworm's own mucus,
remains damp enough for the earthworm to chill out for a few
weeks until the heavens open once more.

Bats

Summer evenings teem with bats. Resident in both urban and rural environments, bats are at their most active between May and September. Head out at sunrise or sunset, or just after dark, in dry weather and you'll catch a glimpse of colonies on the wing, dive-bombing and sky-twisting like flying aces. Delicate, unassuming but much maligned mammals, bats are a key part of our nocturnal environment, keeping huge numbers of insects such as midges and mosquitoes in check while bird species are snoozing. Worldwide there are around 1,400 species of bat, with 18 in Britain alone.

The bat is most at home in one of three places: nooks and crannies in buildings, underground sites such as caves and tunnels, and hollows in trees. Only two British species – the greater and lesser horseshoe bat – roost upside down. Small bats such as the common pipistrelle prefer to crawl into tight, cozy spaces behind window frames and under roof tiles. Larger bats, such as the brown long-eared bat, need roomier accommodation such as large roof voids or barn spaces.

Wherever bats roost, they need to be near a source of food. Private gardens and wooded parkland, both urban and rural, can provide plenty of insects for species such as the common pipistrelle or

serotine bat. The latter is often found flitting around lampposts, hoping to snaffle moths mesmerized by the light. Other bats, such as the soprano pipistrelle or rare Daubenton's bat, prefer wetland habitats. Canals, ponds, lakes and slow-moving rivers all attract these night-time patrollers on the hunt for flying insects. Others, such as the Barbastelle and Bechstein's bat, are woodland specialists and make their homes in mature, veteran trees – a habitat that is constantly under threat. One of the least likely places to catch a glimpse of these night-time aviators is wide-open land. Bats prefer to stick to places with cover to avoid becoming a bedtime snack for owls and domestic cats, two of their most energetic predators.

Many of our bat species have declined over the past few decades. There are, however, volunteer groups who are championing these sublime creatures and organize evening bat walks during summer. At the end of August every year there is also an International Bat Night, held across 30 countries, which is designed to both celebrate and educate people about bats and bat conservation work. Why not flock to your nearest event and experience bats in their natural environment, as they flit and swirl in the darkening sky?

Wasps & hoverflies

Wasps need a public relations officer. For a creature so central to a healthy ecosystem, it's spectacularly unpopular. And yet, what a marvellous insect it is. With over 100,000 different wasps known to science, and 7,000 different species in Britain alone, this striped crusader is a hero of the ecosystem. Not only is it an energetic pest-killer, seeking out bothersome insects such as flies and aphids, the wasp is also a key pollinator of crops and flowers. Indeed, for some plants, only the wasp will do. Many fig trees, for example, are exclusively pollinated by fig wasps, while dozens of species of orchid rely solely on the wasp for survival. The wasp is also thought to play a special role in maintaining and spreading one of the human race's most prized natural wonders – brewer's yeast. Recent research has shown that certain wasps store this remarkable yeast in their guts throughout the year, spreading it from grape to grape when they buzz through a vineyard.

One of the most common wasps – the yellowjacket – pulls no punches about advertising its sting. Its vivid black-and-acid-yellow bands are a clear signal to predators to stay well away. An altogether more timid insect, however, has copied the wasp's clothing. In a form of self-defence called Batesian mimicry, the hoverfly has become a master of copy-cat disguise. Different species of hoverfly have evolved to look, at first glance, like aggressive wasps, bees and hornets. This clever masquerade fools birds and other predators into keeping their distance. But take a closer look and you'll soon see through their camouflage. Unlike wasps and bees, which have two sets of wings, the hoverfly zips about on just one pair and has no pinched waist like a Victorian corset. You'll also notice the hoverfly has short antennae and huge eyes like aviator sunglasses. For all its bravado, the hoverfly is a big softie and can neither bite nor sting. In fact, underneath its confusing outfit, the hoverfly is one of the world's most useful creatures and, after wild bees, our second-most important pollinator.

Moths

Show-offs always get the attention. Compared to their glamorous butterfly cousins, moths are one of the least appreciated of all nature's creatures. Across the world, there are thought to be around 165,000 different species of moth compared to just 18,000 butterflies. In Britain alone, while we gaze in admiration at the 60 or so colourful butterflies that visit our gardens and parks, we happily ignore the 2,500 species of moth that flutter in the shadows.

Moths are often thought of as dowdy. It's hard to get excited about a mousy-coloured insect compared with the showmanship of a peacock butterfly. And yet, moths are canny in their invisibility. Many species fly at night, and rest in the day, making them vulnerable to predators. Cloaking yourself in bark browns and lichen greens is a smart move if you don't want to get eaten.

And not all moths are washed in the colours of concealment. Despite their dull reputation, many species of moth are dazzlingly good-looking. From the spotted dalmatian ermine moth, with its brilliant white fur, to the deeply impressive lime hawk-moth, there is a painterly, refined beauty to many different species. Some are also as bright as butterflies – the six-spot burnet and garden tiger moths, for example, certainly give a tortoiseshell a run for its money. While learned botanists obsessed over butterfly classification, it was often left to ordinary rural folk to name the moths they found in their fields and hedgerows. Many have delightful monikers such as the true lover's knot, foxglove pug, heart-and-dart or ghost moth. One species, Mother Shipton's moth, is even named after a soothsaying witch whose crooked silhouette haunts its forewings.

While people today can still recognize a handful of butterflies, the names and descriptions of moths now elude us. Many are also unconfident about what separates a moth from a butterfly. Broadly speaking, moths tend to come out at night, have comb-like or feathery antennae and rest with their wings open. Butterflies usually sit, wings closed together, have straight or clubbed antennae and zip around during daylight hours. In reality, the distinction between the two is often blurred and many species don't obey the rules. As moths have been traditionally under-studied, less is also known about their distribution and behaviour. Citizen science, therefore, has a critical role to play.

Moths are almost everywhere, not just in rural locations. From back gardens, to public parks, scrubland, and waste ground, there are often dozens of species of moth living in suburbs and cities. Every summer across Europe, there are National Moth Nights where people are invited to go out and record moths, at home or at an organized event, to help celebrate these bewitching creatures and monitor their progress. You can also contribute to ongoing, year-round schemes such as the National Moth Recording Scheme (NMRS) which has, to date, collected over 34 million sightings of moths across the UK. Whether it's a caterpillar, silky cocoon or adult moth, every sighting is precious.

Dragonflies & damselflies

Nature is the mistress of engineering, and few aeronautic creations are as sublime as the dragonfly. It darts and dives, hovers and accelerates, commanding the skies with precision. Almost defying the laws of flight, the dragonfly effortlessly soars up and down, left to right, forwards and even backwards. Such agility comes from the dragonfly's ability to control each of its four wings independently. A sharp turn, a graceful hover, a rapid acceleration – nothing is too taxing for this airborne acrobat.

And yet, for centuries, humans reviled the dragonfly. Fear, perhaps, or envy attracted unkind labels for this harmless creature. Many thought it a flying snake and called it the adder bolt, flying asp, penny adder and other reptilian misnomers. Even its current name comes from the original meaning of the word dragon as 'serpent'. Others feared it stung or harmed livestock. Colloquial names across Europe include the rather alarming Devil's needle, mule stinger and big spike. Only one name approached any kind of begrudging appreciation – libella, or little balance – a nod to the dragonfly's aerial stability.

The damselfly, the dragonfly's petite relative, was luckier. Damsel comes from *demoiselle,* the French for young lady, a name perhaps gifted for the insect's graceful movements and delicate frame. It was also sometimes known as the water butterfly, a name that recognized the damselfly's butterfly-like fluttering and habit of resting with its wings closed.

Both dragonflies and damselflies spend most of their lives as larvae in water, but in summer they emerge as adults ready to fly, eat and mate. It's a short life but a merry one – small damselflies rarely live longer than a fortnight on the wing, while dragonflies might last the whole summer before chilly weather takes its toll. A warm summer's day is the optimum time see both dragonflies and damselflies circling ponds, rivers and wetland nature reserves looking for food and a chance to mate. Several species of dragonfly are also confident long-haul flyers and will head to woodland rides, hedgerows and moorland to search for insects.

Grasshoppers & crickets

Hot weather incites the see-sawing concerto of the cricket and grasshopper. The terms are often used interchangeably, but there are easy ways to tell our violinists apart. The grasshopper, as the name suggests, tends to stick to grassland, while the cricket is found almost anywhere. The names also give a clue to their differing diets. Grasshoppers are vegetarians, while crickets are omnivores, slurping down a wide range of foodstuffs from aphids to fruit and flowers. Grasshoppers have short, stiff antennae; crickets long, wavy, stylish ones. Grasshoppers are active in the daytime, while crickets get into their stride at dusk. Even how they make their music differs. To create that characteristic staccato, grasshoppers stroke their back legs against their wings while crickets rub their wings together. Either way, the chirruping, high-pitched performance is often a love song sent floating out into the balmy air. Pick me. Pick me.

Cloud spotting

For hundreds of years, people thought of clouds as unclassifiable. Each was so unique and temporary, it was believed that no meaningful similarities could be made between them. And, while it's still true that no two clouds are the same, we have the genius of one eighteenth-century sky spotter – Luke Howard – to thank for creating a system we still use today.

Howard had spent his childhood looking upwards, analyzing weather patterns and measuring meteorological phenomena. Based on his observations, he came up with three fundamental cloud shapes – *cumulus* (the Latin word for clump or heap), *stratus* (flat or layer) and *cirrus* (curl of hair). Starting with these basic building blocks, Howard could also describe different versions of these clouds based on where they sat in the sky, *altostratus* (alto, meaning high) or intermediary forms that seemed to combine two types of cloud, e.g. *stratocumulus* (a layer of clumpy cloud).

Summer, with its balmy and often unsettled weather, is a great time to cloud-spot. And, while we often associate clouds with wet, rainy weather, some kinds of cloud only form on sunny days or are at their most dramatic when the temperature is warm. They can also indicate what kind of weather to expect. Here are some of the main types of cloud to look out for at different altitudes.

Low clouds (Below 1,800m/6,000ft)

Cumulus – often described as cotton wool-like, these puffy, white individual clouds usually form on a sunny summer's day and indicate fair, stable weather.

Stratocumulus – a layer of cloud that resembles lots of cumulus clouds joined together. Although often grey in colour, these clouds rarely turn into rain.

Stratus – featureless grey blanket that covers the sky and can hide the position of the sun. Forecast is dull or foggy but usually results in only a light drizzle.

Mid-level clouds (1,800–5,500m/6,000–18,000ft)

Altocumulus – a sky full of tightly grouped white cumulus, like a field of sheep. Indicates fair weather with a possible thunderstorm or dramatic sunset to follow.

Altostratus – nicknamed the 'boring cloud', this is a plain grey sky that's gloomy but doesn't obscure the sun. Often forms ahead of a rainy front.

High clouds (above 5500m/above 18,000ft)

Cirrus – wispy, feathery strands of white hair. Cirrus clouds also take on the colour of a russet or golden sunset. Usually seen on calm, blue-sky days.

Cirrocumulus – called 'mackerel skies' because these multiple, rippling white cloudlets resemble fish scales. Fair-weather clouds that can prelude a storm.

Cumulonimbus – a high-rise tower that stretches from low altitude to as high as 9,000m (30,000ft), these vast, dense clouds are associated with extreme weather and produce heavy thunderstorms, lightning, severe gusts and hail. When a cumulonimbus cloud gets really big, it becomes an 'anvil cloud', so named because its top flattens and spreads out to create a dramatic anvil shape.

Herping

Reptiles and amphibians share something in common. They're both ectotherms – or 'cold blooded' – which means they need an external source of heat to stay warm. Their body temperatures also fluctuate with the environment. In very cold weather, both reptiles and amphibians often go into brumation – a kind of short-term sleep – to avoid the worst excesses of the weather and conserve energy. But the warmth of the summer encourages both reptiles and amphibians to get out and about, just at the time when the animals that they eat – such as insects, molluscs, or small mammals – are also at large. Temperature even affects how quickly some amphibians grow. Warmer ponds encourage frog, toad and newt tadpoles to develop and turn into adults more quickly than cool ponds. Because amphibians need moist skin to breathe, however, once they've emerged onto land they're very vulnerable to drying out in hot weather. It's a delicate balancing act.

Summer is often the time of year when you're most likely to spot reptiles and amphibians, although at different times of the day. Froglets, toadlets and baby newts (called efts) leave their ponds and head for nearby, secluded areas to keep moist. These amphibians tend to be nocturnal, coming out on damp nights to hunt for food, and spend a clandestine daytime under logs, stones or other dark spaces. Reptiles, by contrast, occupy different temporal niches depending on the species. Slow worms, for example, absorb heat by hiding under warm places during the day – such as compost heaps – and then combing for food at dusk. Grass snakes, by contrast, prefer basking in the sun and tend to hunt during the day, hanging out by ponds and wetlands waiting for their favourite prey – amphibians.

'Herping', or seeking out reptiles and amphibians, has become increasingly popular but both types of creature are acutely sensitive to disturbance. If you want to hop onboard, work with one of the many specialist conservation bodies instead, which welcome volunteers for herp-friendly activities, including habitat restoration, reintroductions and surveying.

Harvest mouse

When scientists tried to name the harvest mouse, almost nothing could convey its littleness. Europe's smallest rodent, the harvest mouse weighs barely 5g (⅕oz), about a sixth of its house mouse cousin. In the end, they plumped for *Micromys minutus* – literally 'teeny tiny mouse' – a description that conveys its adorable size but nothing of its other remarkable talents.

The harvest mouse is an acrobat. A life spent among tall grasses and cereal crops has created a tiny rooftopper rodent capable of leaping, climbing, and gripping onto slender stems with ease. Its prehensile tail – a rare superpower – also helps it to hang on while foraging for food and building its nests. Its name gives away its habitat of wheat and oat fields but the harvest mouse also makes its home in other places with lofty plants. From roadside verges to reed beds, the harvest mouse plumps for places with plenty of seeds and insects. It also needs the materials to construct its high-rise nest, a hollow cricketball of woven grass suspended at knee-height off the ground.

Late summer was often the time when country folk traditionally came face to face with the harvest mouse. Come scything time, these minute mice and their nests would be accidentally gathered in along with the sheaves. Harvest mice would find themselves overwintering in haystacks or with loose straw stored in barns, often with plenty of grain on hand. With changes to farming practice, such as earlier ripening crops and the use of thundering combine harvesters, the harvest mouse has largely abandoned agricultural land. It's now a creature more likely to be found in places with tall grasses and reeds, such as rough tussocky grassland, meadows, marshes and field verges. And, although nocturnal for most of the year, late summer is probably the optimum time to see a harvest mouse swinging about in the daytime, dangling from a grass stem by its grippy, daredevil tail.

Daisies & dandelions

The humble daisy was the medieval scholar's *solis oculus,* or 'sun's eye'. But everyday folk had thought of a better name, *dæges eage* – 'day's eye' – from the petals' habit of opening at dawn and closing at dusk. It was also a flower most often picked for love oracles – the 'he loves me, he loves me not' game played by hopeless romantics. Luck was on their side, however, as most daisies have uneven numbers of petals – 'he loves me' usually wins. The daisy's showier sister, the dandelion, was also plucked by the lonesome and lovelorn. Once transformed into a clock, the dandelion could predict your fate. From marriage to money, babies to lifespan, the puff that removed the very last seeds revealed the one true answer. The real love affair, however, is between these 'weeds' and pollinating insects. Both species are hugely important sources of nectar and pollen for some of our most treasured bees, butterflies and beetles.

Meadows & wildflower seeds

The landscape is a conversation between people, animals and their environment. Few natural spaces are truly untouched wilderness. Instead, many of the places we traditionally think of as countryside are the result of an ongoing dialogue between farming and wildlife. A prime example of this are meadows. The word comes from the Anglo-Saxon word *mædwe*, meaning 'field mown for hay', and reveals the practical purpose of this most cherished but fast-disappearing rural feature.

Meadows have a very specific job. They are areas of grassland that are allowed to grow ungrazed between spring and late summer. Historically they were cut in late August by scythe, and the hay collected and stored for winter to feed livestock. During autumn, sheep and cows lightly grazed the stubble but come spring, the meadow was again off limits and left to flourish once more.

This ancient type of land management produces a very particular and species-rich environment. Because the wildflowers and grasses are allowed to ripen in late summer, and go to seed, they bloom year after year. Cutting the hay just as it goes to seed means it's full of goodness for livestock and, crucially, prevents it decomposing and making the soil too nutrient-rich for anything but grass. Scything the meadows, although slow and labour-intensive, also allowed wildlife living among the stalks to escape or take refuge. Even the act of letting sheep and cows nibble the stubble helped the meadow to thrive, their cloven hooves pressing the wildflower seeds into wet soil.

The sheer diversity of wildflowers in a meadow encourages a glorious array of invertebrates, birds, mammals and other creatures. From brown hares to toads, harvest mice to skylarks, the unique meadow environment supports many of our most treasured countryside residents. Some of the most common wildflowers in meadows also support an astonishing number of animals. The sunshine yellow bird's foot trefoil, for example – also known as eggs and bacon – provides food for at least 130 different species of invertebrate including the caterpillars of rare moths and butterflies, and many different wild bees.

Few of us have the land or resources to create our own large wildflower meadow. But we can easily transform a border, window box or small corner of the garden into a haven filled with grasses and pollinator-friendly wildflowers. Wildflower seed mixes, seed bombs, and even wildflower turf can bring life and colour to small areas of bare, weed-free soil. Lawn competes with native wildflowers so it's important to strip off any layer of grass before sowing. Wildflowers also tend to be sun-worshippers and prefer well-drained soil, although some species tolerate light shade, dampness or coastal sites. Choose a mix of wildflowers that suits not only your ground conditions but your location. In late August or early September, after your wildflower patch has finished its display, cut it back to ankle-height and shake out any remaining seeds before taking away the debris. That way you'll have another summer of blooms, butterflies and bees to look forward to.

Autumn

In 1841, the writer George Eliot was penning a letter to her dear friend, Miss Lewis. 'Is not this a true autumn day?' she asked rhetorically. 'Just the still melancholy that I love – that makes life and nature harmonize. The birds are consulting about their migrations, the trees are putting on the hectic or the pallid hues of decay [...] Delicious autumn! My very soul is wedded to it, and if I were a bird I would fly about the earth seeking the successive autumns.'

Eliot clearly adored autumn. And she also recognized its inherent poignancy. For many plants, autumn is the point of no return, the time of year when growth tips over into slow decline. Heads turn to seeds, stems stiffen and leaves begin to change colour. Creatures, too, prepare to face the tougher half of the year. Many will spend the season fattening up to ride out winter, or pack their bags and head for warmer shores. Even the weather feels like it's on the turn; while early September can steal a few extra days of sunshine, there's no escaping the season's slide into longer, chillier nights.

Not all is lost, however. Autumn is also famously the season of mellow fruitfulness, when hedgerows and orchards ripen, roses turn to hips and nut trees fill their branches. There's a bounty on offer for those in the know and, for many creatures, autumn is a trolley-dash. Squirrels stash acorns, hazelnuts and fungi, hedgehogs and badgers gorge on earthworms, and garden birds fight over berry-laden bushes.

While many species take a break from the summer of endless courtship, a handful leave their mating until autumn. This is the prime time for deer rutting and the woodlands crackle with testosterone. Solitary wild boars also start to snuffle out a mate. House spiders get together in early autumn and briefly share a nuptial web. Even ladybirds seek each other out for companionable, platonic warmth and head indoors for shelter.

Autumn's most glorious feature, however, is its kaleidoscope of leaf colours. It puts on a fine spectacle, a last hurrah, before everything tumbles to the ground. In French, Italian and Spanish, the words for autumn are broadly similar but English-speakers in the sixteenth century had two names to choose from – 'autumn' and 'fall of the leaf'. When early settlers landed in North America, they brought the 'fall of the leaf' phrase with them, and its shortened version – fall – stuck. Back in Britain, this sweetly poetic description also fell – out of common use – and into the history books.

Despite its cooler weather and shorter days, many people choose autumn and its peculiar beauty as their favourite month. Many late garden flowers are still ablaze but don't demand the constant attentions of midsummer. Deciduous trees, background for most of the year, suddenly pull focus with their rich hues; even stripped of leaves, their architecture still catches the eye. And the first frosts are welcomed in, a gentle reminder that, in a few months, things are about to get a lot, lot colder.

Susurrus

For most of the year, we don't really notice the sound of the wind as it kisses the trees. In autumn, however, as the leaves lose their moisture, the gentle rustling and soughing of the season's breezes soon become apparent. One of the most evocative words to describe this whispering sound is the gorgeously onomatopoeic 'susurrus'. It's a surprisingly useful word – and can mean any kind of soft murmuring or humming sound – it was historically applied to everything from tinnitus to a babbling brook.

The ancient Romans particularly loved the word and used it freely. From shared secrets to the sound of a contended beehive, susurrus was the perfect descriptor of whispers and white noise. It was also used as a shorthand for a secret adviser, an official who poured confidences into the ear of an emperor.

While the word's popularity has waned over the years, susurrus still perfectly describes the gentle swishing of autumn leaves. It's a sound unlike anything else. Listen closely and you can hear that it's a softly spoken, almost conspiratorial conversation between the woodland and the wind.

Leaves

During autumn, the trees are set aflame. Leaves that were intense green burst into a fiery palette of yellows, oranges, reds and russet browns. Then, as the ambers of autumn die down, this foliage finally separates from its parent tree and tumbles to the forest floor. It's a remarkable pageant and yet few of us know why, or how, the colours of autumn change.

During spring and summer, leaves are green. This is because they are full of chlorophyll, the green pigment that allows the tree to turn sunlight into simple sugars. These sugars keep the tree alive, giving it the nourishment it needs to grow and develop. When autumn begins, the shorter days and cooler temperatures tell the tree it's time to stop producing chlorophyll. This allows the leaves' other, hidden colours to show through.

Carotenoids, for example, are always present in leaves but are usually masked by green chlorophyll. These are finally given a chance to shine, giving autumn leaves their yellow or orange hue. Some trees also begin to produce a new group of pigments towards the end of summer called anthocyanins. These are what give certain trees, such as red maples, their dramatic russet, pink and purple leaves. The leaves of some species, such as oak, also turn brown because of a pigment called tannin. This is also always present but only revealed when both green chlorophyll and yellow-orange carotenoids have faded.

The old wives' tale that rain washes the colour from autumn leaves does have a grain of truth. The weather can affect just how impressive an autumn display we'll get, at least when it comes to producing anthocyanins. For a truly magnificent show, the best seasonal conditions are mild sunny days followed by cool nights. Cloudy or rainy days limit the amount of red colour a leaf produces, while heavy gusts and storms strip a tree's leaves before the foliage has reached its peak. Jack Frost can also be bad news. Sub-zero temperatures can damage leaves, dampening down their ability to turn from cool greens to glorious, fiery shades.

House spiders

Have you ever noticed that, come September, house spiders seem to come out of the woodwork? It used to be thought that the cooler weather brought eight-legged interlopers indoors but it now seems that these secretive creatures were probably in your home all along. September is the prime month for house spider mating. While females stay in their cobwebs, usually well hidden from view, the males venture out and about in the hope of finding a companion. If you see a house spider scuttling along the carpet, it's probably a hopeful male suitor, scouring your abode for a partner. If he's lucky and finds a lover, she'll let him move in, and between now and winter they'll co-habit in her cobweb. It's a brief honeymoon, however, as the male will die, exhausted, before winter, leaving the female to hatch their handful of young the following spring.

Hedgerows

In September and October, the hedgerows hang heavy with wild berries. It's one of nature's delicious tricks, to disguise a plant's seeds in a fruity coating so birds and mammals devour and disperse them. Hedgerows are also a lifeline for native wildlife, often a last chance to line the belly before winter depravations set in.

We humans have also long relied on hedgerows. Many of the fruits and berries of the wayside are packed with nutrients. Both rosehips and crab apples are so rich in vitamin C, for example, that during World War II the government turned to hedgerows to boost the nation's health. Our relationship with hedgerows, however, has even deeper roots. For thousands of years they were used as living boundaries to keep both livestock in and wild predators out. Both 'hedge' and 'haw' mean the same thing in Anglo-Saxon – 'enclosure'. Hawthorn, the most common of the hedgerow plants, literally means 'thorny hedge'.

The value of a hedgerow to our ancestors was so great it was codified in law. Hedgerows not only provided a free larder of wild food, but also firewood and timber for tools and furniture. The common rights of *hedgebote* and *haybote* are ancient and allowed peasants to take hedgerow timber to fix fences and make implements for haymaking, such as rakes and forks. Hedgerow plants also wound their way into medicines and folk cures. Cleavers or sticky-weed, which was known as 'hedge-rife' in Anglo-Saxon times, was used widely in cures for everything from earache to epilepsy. Another hedgerow herb, bryony, also known as wild neep, is a dangerously effective emetic and laxative, a quality highly valued by medieval doctors and home herbalists alike.

During early autumn, the hedgerow is laden with edible pleasures. Come September, most of the blackberries and wild raspberries are over but the few that are left are at their absolute sweetest. Hawthorn berries, which are also known simply as haws, are at their peak and make delicious vinegars and ketchup. Haws are also stuffed with pectin, the ingredient that helps jam to set and so are excellent for adding to jellies, conserves and fruit leather. Seek out rosehips too – they make floral, tangy syrups and cordials, as well as jellies and jams. Even rowan berries, a fruit most of us ignore, can be transformed into a perfumed, slightly bitter jelly that once enlivened medieval game feasts.

And don't forget the humble sloe, the fruit of the blackthorn. The name probably comes from an ancient root word meaning 'bluish' and the fruits do, indeed, look like temptingly tiny plums. There's a sting in the tail, however, as sloes are mouth-puckeringly bitter and blackthorn branches have fierce spikes. Traditionally, rural folk would wait until the first frost had nipped the sloes, or 'bletted' them, which softened the fruit and removed some of its acidity. Only then would you add the sloes to your spirit of choice, seal up the bottle and wait until the depths of winter to enjoy its inky sweetness.

Acorns & conkers

'Potential' – such an interesting word. We use it now to mean 'capable of becoming' but it comes from the ancient Latin *potentia*, meaning 'formidable power'. And few words could better be used to describe autumn's acorns and conkers. While the acorn is the nut of the mighty oak, the glossy conker belongs to the horse chestnut tree. Both seeds contain all the information and starter-fuel needed to grow into gigantic, mature trees. Horse chestnuts can live for three centuries, their crowns reaching heights of 30m (100ft) or more. While oak trees rarely reach much higher, their capacity for longevity is extraordinary. Thanks to medieval monarchy's obsession with hunting forests, England has more ancient oaks than all the other European countries put together.

The girth measurement gives away an oak tree's true age; a trunk 5.5m (18ft) in circumference suggests it was just an acorn around the time of the Civil War, in the middle of the seventeenth century. An 8m (26ft) girth and the acorn belongs to the age of the Black Death, the beginning of the Renaissance and Robert the Bruce. And an oak tree with a bulging waistline of 9m (30ft) or more is a truly rare survivor, a likely witness to William the Conqueror's Battle of Hastings and the end of Anglo-Saxon rule.

To reach this venerable age, an acorn or conker must have luck on its side. Of the thousands that tumble to the ground from each tree, only a handful germinate and grow. Experiments have shown that of a relatively small oak tree's 14,000 fallen acorns, only 20 will successfully germinate and turn into healthy seedlings. Most will be eaten by animals or infested by insects, while the rest of the acorns will be sterile or fail to thrive as seedlings. And even if an acorn or conker does manage to sprout and grow into a tiny sapling, it still has decades of hard winters, dry summers and raging winds to contend with. It's a miracle that any tree reaches a century. To survive a thousand years, from tiny seed to corpulent, gnarly veteran, is truly against all the odds.

Crab apples

We have a lot to thank the crab apple for. All apples, from the Granny Smith to the Golden Delicious, belong to the same genus – *Malus*. Alongside these sweet, orchard varieties sit their sour cousin – the wild crab apple – a fruit tree with a wide distribution across the northern hemisphere, including Britain, Europe and North America. Different species of crab apple are native to different regions – South Asia has its own *Malus orientalis*, for example, while Britain and other European countries enjoy their own sharp-tasting variety, the European crab *Malus sylvestris*. Modern tastes shun the crab apple but our ancestors knew its worth. For thousands of years, people picked and dried crab apples in early autumn, concentrating their sweetness and preserving them for the long winter. Our forebears also made verjuice from crab apples, a sour condiment used a bit like vinegar. From salad dressings to rich sauces, verjuice not only enlivened many a dish but was also packed with health-giving vitamins.

All modern orchard apples are descended from one species of crab apple – *Malus sieversii* – which evolved in the Tian Shan mountains in Central Asia. From small beginnings, the fruit grew from a bitter crab into a sweeter, juicier apple, one that caught the eye of traders travelling along the Silk Route and slowly wound its way across the Mediterranean, finally reaching Britain around the time of the Romans. Along the way, this plump, new apple hybridized with native crab apples from each local region, producing a huge number of varieties. Look into the genetics of a supermarket apple and you'll find an exotic heritage – about 50 per cent of its DNA comes from the Tian Shan apple, around 20 per cent from the European crab and the last 30 per cent is a ragtag assortment of unknown crabs as yet to be identified.

The crab apple is also a tree beloved by wildlife, especially during early autumn. As the fruit appears, so too do the birds and mammals that relish them. From fieldfares to foxes, blackbirds to badgers, many creatures bolster their diet with the native crab apple. The crab is also one of the few trees that will host the pernickety mistletoe. Come spring and summer, the leaves of the crab apple tree feed moth caterpillars, while its blossom provides nectar and pollen for a multitude of pollinating insects. Plant a crab apple tree and you'll be doing both nature and apple-growers a huge favour. Not only do these ancient, native trees bring wildlife to your doorstep, but crab apples also help to pollinate delicious orchard apples nearby. Most cultivated apple varieties need the pollen from a different apple tree to bear fruit. Because crab apples are in blossom for such a long time, they act as excellent pollination partners for orchard apple trees, many of which have short and varied flowering times of their own.

Red deer (rutting)

At this time of year red deer stags fight for the right to mate. It's
a fierce contest, and keenly fought, with plenty of bravado and a
whiff of danger. As the females watch from their harem, the male
bellows into the mist; a bull's throaty roar sounded over and over,
to scare off potential rivals. His deep-throated groans might be
warning enough but, chances are, he'll have to battle for supremacy,
locking horns with other stags to keep them at bay. Before they make
contact, two stags pace around each other, like sumo-wrestlers full
of bluster. When the moment strikes, it's heads down for a shoving
contest, each attempting to outmatch the other. For three or so weeks
this hormone-fuelled tournament rages across forests and large
parklands. Come February, however, the stags shed their antlers and
grow a whole new set before the following autumn, when the battle
commences once more.

Hedgehogs

Autumn is a race against time for the bristling, rambling hedgehog. While summer was all about finding a mate and raising a family, autumn is the season for self-imposed gluttony. With winter providing few opportunities for food, hedgehogs must use these three months to put on enough body fat to see them through until next spring. In September, finding earthworms, beetles and other insects isn't too taxing – the countryside is still flushed with late-season colour and life – but by early November the pickings can be thin on the ground, especially if the frost shows its face.

Apart from the few weeks a baby hoglet spends in its mother's nest, the hedgehog's life is a lonely one. Most spend their days napping in makeshift, untidy nests of leaves and nights wandering endlessly in search of food. When they do find a juicy worm or crunchy beetle they tuck in with relish. Indeed, a hedgehog's poor table manners are probably the first thing you'll notice. It's often easier to hear a hedgehog than see one, the unmistakable sound of open-mouthed chewing and wet slurping emanating from underneath a hedgerow.

They're sensible to be greedy. In November, depending on the outside temperature, hedgehogs begin to go into hibernation and must survive on stored fat reserves until the following spring. Even the plumpest of hogs will go on a radical diet over this lean time, emerging in March just over half their autumn weight. It's a perilous plan of action, but a wise one. Almost everything that the hedgehog loves to scoff – from invertebrates to birds' eggs – either hides or doesn't exist over winter, making finding a meal more hassle than it's worth. Hedgehogs are also terrible at keeping themselves snuggly and warm. Somewhere in their evolutionary past, the hedgehog swapped its fur for spines, a development that prioritized defence over deep insulation, leaving it very vulnerable to the cold.

Creating useful piles of plant debris, in quiet corners of your garden, gives the hedgehog a fighting chance. At this time of year, a hedgehog starts building its hibernaculum and will take material from these heaps. More substantial than a loose nest, its winter home is tightly woven with twigs, grasses, leaves, bracken and other plant material and keeps the hedgehog warm through the worst of the weather. It also chooses a suitably sheltered spot – at the base of a hedge, in an old burrow, or even under a garden shed. Piles of plant debris shelter insects escaping the cold too, which provide a ravenous hedgehog with a welcome snack. Even better, leave a small saucer of dog or cat food out at night-time, in a home-made hedgehog feeding station, to give our prickly friends a hearty meal before they hunker down. Only with a full tummy will they happily curl up in their hibernaculum, close their eyes and drift off into their seasonal slumber.

Red squirrels

Autumn is a busy time for the industrious red squirrel. The mast of the forest – the nuts, cones, seeds and berries – is ready to pick, and will sustain these tufted, russet squirrels over winter. Red squirrels take great care to cache this food, burying it in shallow holes in the ground. They are also rather fond of fungi and stash it in tree crevices. This secret hoarding helps the red squirrel survive over winter without needing to hibernate. But squirrels are also forgetful and often fail to rediscover their hidden treasure. This absent-mindedness is a critical part of the woodland ecosystem, helping to disperse fungi and encourage the growth of new trees. In return for this invaluable labour, red squirrels – which are in dramatic decline – need our help. Adopt one through a charitable wildlife organization and you'll help secure the future of both the red squirrel and our rich woodland habitat.

Geese & swans

We often love to complain about chilly weather. But for some species of migratory birds, western Europe's cooler months are positively balmy. Many of the geese that migrate to our shores in autumn are escaping even colder extremes back home. When the temperatures really start to plummet across the Arctic Circle and Nordic countries, geese take to the sky. Some of their journeys are remarkable; dark-bellied brent geese, for example, abandon the Arctic tundra of northern Russia and fly huge distances to reach the temperate comfort of England's estuaries. Their pale-bellied cousins make an equally impressive pilgrimage, departing from Canada and Greenland to make a 4,000km (2,500-mile) beeline for Ireland, Scotland and France.

The diminutive Bewick's swan, another autumn visitor, leaves its Siberian breeding ground and seeks sanctuary in East Anglia, the Severn Estuary and Lancashire. There, it'll hoover up leftover grain and potatoes in farmers' fields during the day and spend its evenings roosting safely on open water. Come spring, migrating geese and swans will once again up sticks and head back to their summer residences to breed.

Some species of geese have decided to turn their autumn holiday destination into a permanent home. The greylag goose, for instance, has two populations in Britain. Although around 100,000 wild greylags arrive from Iceland every autumn, and overwinter in Scotland, lowland UK also has a large population of year-round residents. These stay-at-home geese are probably descended from domesticated populations of greylags and have, over time, lost the instinct to migrate.

One species that rarely migrates into Britain is the Canada goose. For a bird with such a long-distance moniker, the Canada goose was in fact brought here unwillingly. Native to North America, the bird was introduced to Britain in the late seventeenth century. Charles II imported it as a pretty ornament for his waterfowl collection in St James' Park, a fashion statement quickly emulated by other members of the aristocracy. Now well established, the Canada goose is one of the most commonly recognized waterfowl species – with around 190,000 individuals in Britain alone – equally at home in gravel pits and city parks as in remote estuaries and meandering rivers.

Ladybirds

As the days shorten and the outside temperature begins its inexorable descent, the ladybird senses that winter is around the corner. Few of her kind could survive an Arctic snap, but the real problem is the lack of their favourite food – aphids – which overwinter as eggs. And so, wisely, the ladybird calls it a day and heads for a warm, tight crevice to wait it out until spring. Some of her fellow ladybirds will go it alone and crawl into a wall crack or behind loose bark to spend the next few months in splendid isolation. Other species of ladybird prefer safety in numbers and hibernate in clusters, huddling together in window-frame gaps or garden sheds. Few begrudge the ladybird this autumn bed-and-board, however. Thanks to her exemplary aphid-munching skills, she's beloved by farmers and gardeners alike. In fact, folklore tells you that if a ladybird graces your presence, a spot of good fortune is bound to follow.

Toadstools

While most people call fungi 'mushrooms', they reserve a special word for those deadly species that have haunted folklore for centuries: toadstools. These woodland assassins cast a mortal shadow over our benign edible varieties and, such is our fear of them, that the act of collecting wild mushrooms often feels more like a game of Russian roulette than wholesome foraging.

The association between toads and toxic mushrooms is no accident. Both toads and toadstools were long believed to be highly noxious, their secretions capable of great harm – one of the earliest uses of the word comes from the late fourteenth century: 'tadstole'. Witches were thought to use toad poison in their terrible potions and keep them as 'familiars', creatures that carried out evil tasks at a witch's behest. The warty skin of a toad also invited comparison with the skin of certain harmful fungi, including the panther cap and our fairy-tale favourite, the fly agaric. The, abracadabra, appearance of fungi in autumn, as if they'd sprung from the leaf mould of the forest floor, also tricked the incredulous. Some thought that poisonous mushrooms shot up from toad droppings or magically grew from slimy trails left by slugs, snails and amphibians.

While fears about toads were largely unfounded, our ancestors were right to be wary of toadstools. Not only will certain species make you ill or hallucinate if ingested, some are truly lethal. To add to the jeopardy, noxious species can look alarmingly similar to edible ones. The deadly webcap, for example, looks suspiciously similar to an edible chanterelle and often tricks the unwary. So too the panther cap, a speckled brown toadstool often mistaken for the rather delicious blusher mushroom.

Some toadstools helpfully advertise their harmfulness; the fly agaric is unmistakable with its storybook red hat and white spots. Others look innocuous but hide a deadly secret. The death cap, for example, is one of the world's most toxic toadstools. Just nibbling half of one is enough to kill an adult. Another toadstool that lives up to its name is the destroying angel, a pure-white specimen concealing a murderous heart. Delivered the cruellest of deaths, the victim often feels sick initially but then briefly rallies for a day or so, only to slip into a deep and irretrievable coma.

Toadstools should invite our respect, but we can also celebrate their beauty. The fool's conecap, for example, is a delicate delight, a tiny tan parasol that pokes out from the leaf litter. So too the small, cloche-shaped funeral bell, with its characteristic brown ruff, that colonizes rotting logs on the forest floor. Even the angel's wings, a toadstool that can cause brain damage, is breathtaking with its snow-white trumpets and gills. Seek out toadstools and revel in their good looks. When it comes to foraging, however, maybe leave the risk-taking to the experts.

Wild boar

Autumn is a season of bounty for the wild boar. Creatures of the forest, they will spend the next few months gorging themselves. From beechnuts to acorns, crab apples to chestnuts, wild boars must eat with determined gusto if they're to successfully put on weight and breed over winter. Rootling around on the forest floor, they also dig out roots, earthworms, grubs and other delicious morsels to satisfy their voracious appetites. This constant snuffling and scrabbling plays a unique role in the arboreal ecosystem; wherever they stick their snouts, wild boar rotavate the soil and help disperse woodland seeds. They also dig up brambles and bracken, both of which can outcompete other native plants and bulbs.

Wild boar live in social groups called sounders. A formidable matriarchal household, the sounder consists of sows and their piglets, or 'humbugs', named after their ginger-brown stripes. Very young males are welcome too but, at around a year old, they have to leave the sounder and form their own boys' club called a bachelor group. There, they'll stay together for another three years, after which time they're ready to go it alone. For the rest of his days, the mature male boar lives a solitary life, only pairing up briefly between late October and February to mate.

Although wild boar are common across much of the European mainland, Britain has been cavalier with its own populations over the centuries. A favourite quarry of both kings and commoners, the wild boar was all but extinguished by the thirteenth century. Subsequent monarchs attempted to re-establish them for quarry, but it wasn't until the twentieth century that wild boar returned to the country in any numbers, largely thanks to fugitives from zoos and pig farms. Estimates put the current population at anywhere between 1,000 and 5,000 wild pigs, although no one really knows the true figure.

Wild boar are also elusive and largely nocturnal, making it unlikely you'll see one out in the daytime. But, you may just come across evidence of their movements on a forest walk. Large areas of disturbed soil or uprooted vegetation on the woodland floor are a tell-tale indicator that they've been searching for food. Wild boar also have distinctive footprints. Unlike deer, boar leave two small indentations – prints from their dew-claws at the bottom of their hoof. Wild boar poo is also very characteristic – lumpy and sausage-sized, like brown marbles stuck together. As with all pigs, wild boar also love nothing more than a mud bath. If you're lucky enough to stumble upon a shallow mucky puddle filled with footprints, or see a tree trunk splattered in mud at pig height, there's a good chance a wild boar has recently enjoyed a wallow and a scratch before trotting off into the brushwood.

Ferns

If you want to time-travel, look no further than a woodland fern. These incredible plants are spectacularly old. Evolving more than 300 million years ago, ferns watched dinosaurs and mass extinctions come and go, quietly flourishing while the world boiled and froze. Their success lay partly in the ingenious way they reproduce. Ferns don't flower, they send out spores. These tiny particles can float huge distances on the wind, and remain dormant for decades, giving the fern a fighting chance in a chaotic landscape. Some species of fern evolved to be evergreen, giving life and shape to the forest floor even in the depths of autumn and winter. Others learned to slowly die back, only to spring back the following year. As their new fronds unfurl, the tips of these ferns resemble the head of a violin. No wonder these extraordinary plants earned the folk names 'fiddleheads' and 'croziers', after a bishop's curved crook.

Dark skies

Autumn is a stargazer's dream. Throughout summer, the nights rarely get dark enough to put on a sparkling show. Come October, however, and the long but still-mild evenings create the optimum conditions to scour the evening sky. Artificial light created by streetlights and cities can spoil the fun, so head to one of the many official dark sky places or to local events. Reignite your inner navigator and learn how to find true north. Look for the brightest star in the northern sky – Polaris or the North Star. If you're not sure you've found it, look for the unmistakable seven stars of the Plough to its left (also called the Big Dipper or Saucepan) and the five-starred 'w' shape of Cassiopeia to its right. Because Polaris seems to stay locked in the sky, while the other stars move their position, it has long been used as a fixed point for navigation and astronomy. Find the North Star and you'll always find your way home.

Murmurations

From November onwards the early evening sky comes alive. About an hour or so before dusk, the heavens fill with clouds of starlings, swooping and stunt-diving in synchronized displays. These boiling, swirling flights are called murmurations. It's a curious word, and an old one, long used in religious texts to describe blasphemous grumblings or mutterings. Our ancestors must have looked at these shape-shifting, gossipy swarms and thought something devilish about them, but today we appreciate their beauty. Indeed, few sights are as striking. Flocks seems to move as one, changing direction on a whim and staying yet perfectly intact. They rise, plummet, drift apart and coalesce, pulsing like a living organ. And yet, science isn't quite sure how starlings do it or, indeed, why.

Starling murmurations can be vast, with hundreds of thousands of birds moving as one without any particular leader. Studies of starlings suggest that, to remain part of a coordinated flight team, each individual doesn't try to keep up with the entire flock. Instead, each bird responds to information from just seven of its nearest neighbours, a magical mathematical number that allows each starling to maintain its place within a hectic, eddying mass without getting too confused or overwhelmed. Quite what information each starling is paying attention to, or which neighbours it watches, we still don't know.

It's often said that murmurations follow the selfish-herd theory. You flock together for safety in the hope that another member of your group is more likely to be eaten than you. Some scientists are not so

sure, however, and suggest that the large size of a murmuration can actually attract a predator. Instead, the purpose of a murmuration might be the large-scale communication of information, such as food sources and roosting places. Or, murmurations might be a way of confusing predators. The endless curling and diving may dazzle birds of prey, deterring them from attacking for fear of getting injured in a collision or missing their target.

The idea of collective motion fascinates biologists. Not only do starlings exhibit this behaviour, but so too lots of different creatures across the animal kingdom. From clouds of midges to colonies of bacteria, schools of fish to armies of ants, scientists are beginning to understand just how universal and primeval this behaviour is. Even single cells within an organism display collective motion – cells moving together as a group to heal wounds, grow or invade tissues – and science is still working out how they do it.

If you want to see a murmuration, the best displays are to be seen on clear calm days, just before sunset. Starling numbers peak between November and late January. A cosmopolitan bird by nature, huge flocks can be found in places as diverse as woodlands, reedbeds, city centres and coastal resorts. Murmuration chasers can find their nearest potential site at dedicated websites or head to your local nature reserve. And, remember, the next time you look in the sky and see a churning murmuration, you're not just looking at a daring display, but a behaviour that underlines the very building blocks of life.

Winter

Perhaps no other season held such anxiety for our ancestors than winter. For those who were prepared, it was imperative to squirrel away resources, eking them out over three long months until spring reemerged. A poor summer harvest, however, or a grain store ruined, spelled disaster for an entire community, with no other means of feeding themselves. Any person or animal who survived winter could count themselves lucky; indeed, the Anglo-Saxons marked age in 'winters' – a sheep or cow that had seen one winter was an *án-wintre*, two winters a *twinter*. Even being depressed was described in wintery terms – *winter-cearig* could mean either feeling sad because you were getting old or feeling gloomy about winter.

It's perhaps no surprise, then, that winter was often portrayed as a dying man, a human in its final stages of life. Old Man Winter was just one of its names but across time, the coldest season of the year has been imagined as a grey-bearded, withering being. Hunched and shrivelled, barren and pinched, the personification of winter wasn't a sweet, old dear but a stern, hollow-voiced miser. For Shakespeare, the old man was Hiem, a figure who took pleasure in the discomfort of winter: 'hoary-headed frosts, Fall in the fresh lap of the crimson rose, And on old Hiems' thin and icy crown, An odorous chaplet of sweet summer buds, Is, as in mockery, set.' Remnants of these ancient beliefs sprinkle the language of winter. We still describe a certain kind of feathery frost as 'hoary', meaning white-haired with age; cold weather pinches our noses with its gnarled fingers; and snow blankets the landscape with its thick coat.

And yet, for nature lovers, winter can hold so much joy. Few pleasures are sweeter than a landscape, freshly plumped with snow, or a crisp, blue morning glittering with frost. On very cold days, we appreciate many of nature's gifts that we take for granted at other times of the year. The gentle warmth of the sun on a winter's day, for example, even has its own name: 'apricity'. To wrap up in blankets against the freezing cold was once called 'happin', while thick outer clothes for winter earned the name 'haps' or 'hap-warms'. Even the delicious sensation of crouching by a warming fire, to slowly defrost, had its own unique verb: 'to crule'.

Winter is also vital for nature's health. Cycles of freezing and thawing break up soil structure, fluffing it up like a down duvet. Many of our most beloved plants need a period of cold to flourish in spring, too, a requirement called vernalization. From crocuses and daffodils to foxgloves and bluebells, many bulbs and wildflowers require a cold snap to trigger germination. Deciduous fruit trees – apples, crabs, pears and many more – also need a prolonged chill to really bear fruit the following summer. As John Steinbeck felt in the depth of his bones: 'What good is the warmth of summer, without the cold of winter to give it sweetness.'

Flindrikin

Winter is rarely knee-deep in snow. For many parts of Europe and Britain, winter snowfall is sporadic and fleeting, a quick flurry that soon disappears into the cold, damp grey. At times, it can seem like the sky isn't really trying at all. Just a few white flakes swirling to the ground.

This ever-so-insubstantial snowfall has a delightful name: 'flindrikin'. In the Scottish language – a language with an impressive number of words for snow – it refers to a slight smattering. It's not an affectionate word, but not unkind either, and derives its meaning from 'flinder', to move about in a fluttering manner. Applied to both cattle on the loose and dancing butterflies, it perfectly captured the erratic, slightly aimless ramblings of creatures. Creative thinkers also used 'flindrikin' to mean almost anything flimsy. From scanty clothing to frivolous friendships, a measly thin pancake to an empty-headed person, it described them all to a tee.

In other parts of Britian, this sparse snowfall reminded people of ashes. White, carbonized flecks often fall from the sky around a bonfire or open chimney and these cinders were traditionally called 'blenks'. Dialect and folk words can be remarkably poetic – how haunting to describe a midnight flurry as a dark, blenky sky.

Snowflakes

As snowflakes spill and swirl from an opaque sky, have you ever looked up and wondered where they come from? As we already know, there is always water in the air, wafting about as moist, warm vapour. As it rises upwards, the moisture in the air cools and condenses, grabbing onto specks of dust in the atmosphere. As these water droplets grow, they become heavy and plummet back down to the ground as rain.

In winter, however, the sky can become really cold. If the temperature high up in the clouds falls to 0°C (32°F) or less, these tiny droplets of water begin to freeze, creating minuscule flecks of ice. These soon attract more water vapour, which also freezes, making them grow larger and larger. They change their silhouette, too, shape-shifting from simple crystals to hexagons with six arms, or 'dendrites', on each corner. This is the basic shape of all snowflakes.

A snowflake's journey is as important as its destination. Snowflakes get their unique shape because no two fall to the ground in the same way. Different atmospheric conditions affect the growth pattern of a snowflake as it tumbles downwards. A slight change in temperature, air pressure or moisture levels in the air can cause the snowflake to grow in a startlingly different way – sprouting needles here, fern-like fronds there. High humidity, for example, encourages a snowflake to grow at its tips and edges. A snowflake might also collide with another as it falls, snapping off some of its branches or sticking together to create a larger but still uniquely shaped snowflake.

Although all snowflakes start off the same way, and almost always have six sides, the way they change as they make their descent means that no two are alike. Or, perhaps, more accurately, they are very *very* unlikely to be similar. Trillions upon trillions of snowflakes fall to the ground every year and scientists have identified at least 80 different elements that can make up a snowflake, making the possible permutations virtually endless. Some snowflakes are so simple in shape that they look remarkably similar under a microscope. Even these, however, at a deeper molecular level are all slightly different.

And, despite their brilliant white appearance, snowflakes are, in fact, translucent. Light travels in waves and consists of the entire spectrum of colours. When light hits an object, some of these waves are absorbed and some are bounced back towards the eye. A yellow rubber duck, for instance, is yellow because it bounces back the yellow light and absorbs the other colours. When light hits a snowflake, not one but all the colours in the spectrum are reflected back by its ice crystals, making it appear beautifully, mesmerizingly white.

Hibernation

As winter starts to bite, many animals escape its ravages by falling into an extended 'sleep'. There are a number of names for these periods of self-imposed isolation, including torpor, hibernation, brumation, diapause and dormancy, but each has a subtly different meaning. Dormancy – from the Latin *dormire, meaning* 'to sleep' – is a broad term and is used to describe an animal going into a period of inactivity and reducing its metabolic rate. It often does this by slowing its heart rate or dropping its body temperature. This prevents it using up valuable energy keeping warm when resources are scarce and allows it to survive on fat reserves stored in its body.

Hibernation is another word for winter dormancy. (Aestivation is the summer equivalent, when animals hide from extreme heat or drought.) While dormancy can last for weeks or even months, another word – torpor – is often used to describe short bursts of inactivity, often just a few days or weeks. Some animals, such as mice, even use daily torpor as a way to conserve energy in winter, dropping their metabolic and body temperature for just a portion of the day.

Brumation – from the Latin *bruma*, meaning 'snow' or 'wintertime' – is used only in relation to cold-blooded animals such as adders or toads, which hibernate in a slightly different way to warm-bloodied creatures, including waking up periodically to rehydrate. Diapause, on the other hand, is a remarkable strategy employed by a number of creatures – from roe deer to moths – as a response to unseasonably bad weather. Some species can press pause on their growth development or pregnancies if environmental conditions suddenly take a turn for the worse.

Perhaps the most extraordinary adaption to cold weather is known as 'freeze-tolerance'. If a creature can't hunker down and hide from the cold, it may use this remarkable survival strategy. Ice crystals usually damage an animal's cells. But certain species of frog and salamander have developed a way of making their own 'anti-freeze', which is pumped through the bloodstream and allows their bodies to cope with being frozen almost completely solid for weeks. Come spring, and the return of warmer weather, these hardy amphibians simply defrost themselves and hop off into the sunshine.

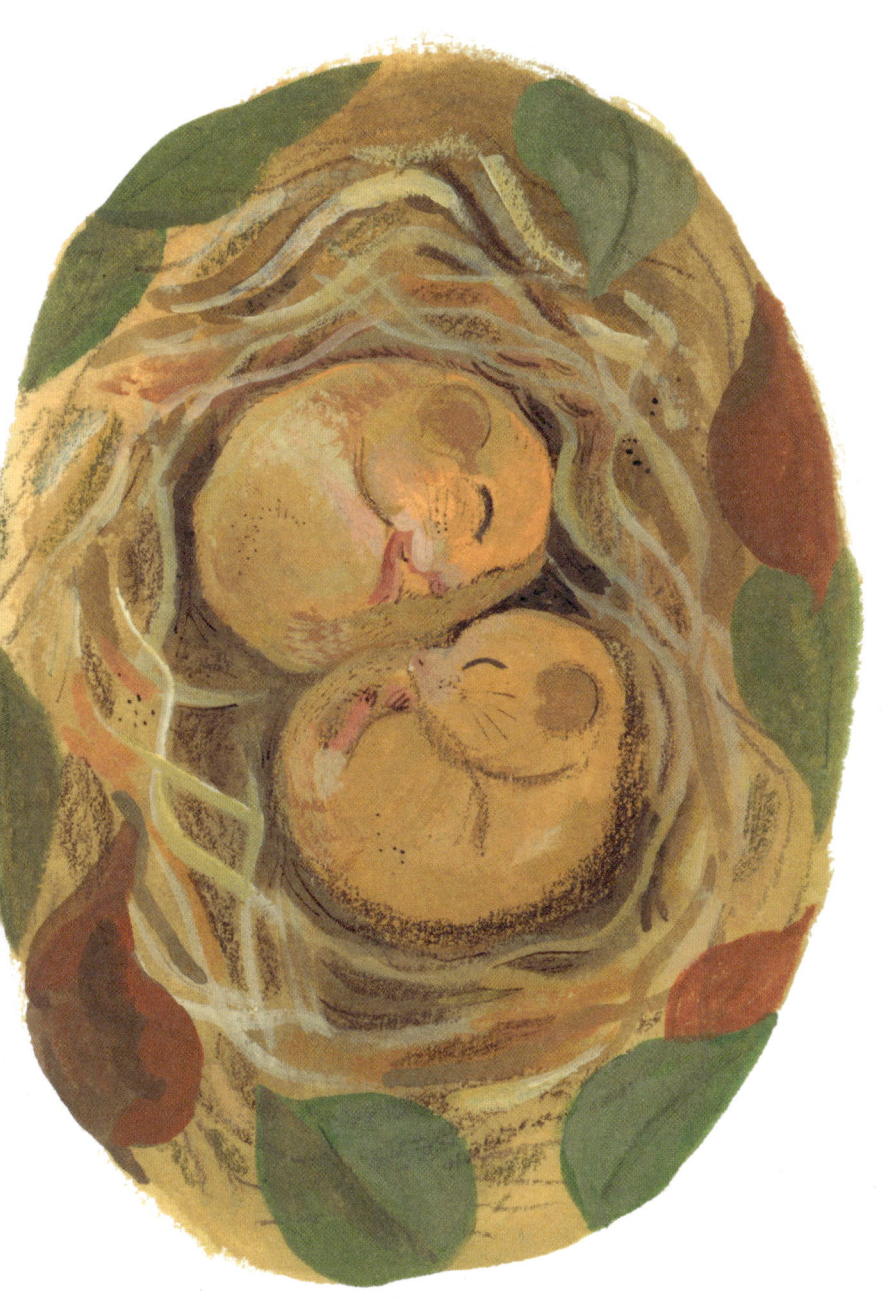

Mistletoe

There is something otherworldly about mistletoe. Unlike other evergreen plants, its feet never need to touch the ground. Instead, mistletoe is a parasite, sucking its nutrients and water from the high branches of unwitting trees. The beginning of this clever theft starts with a greedy bird. Mistletoe seeds are super sticky and will glue themselves to a branch after passing through a bird's stomach or being wiped from a beak. Once adhered, the seed makes itself at home, tricking the bark into growing cells around it and providing everything the young seedling needs to flourish, apart from sunlight. Mistletoe also bewitched our ancestors. For thousands of years, people have been fascinated by mistletoe's grip on some of our most venerated trees, including the apple, lime, hawthorn, and, occasionally, oak. A fulsome bunch of green and pearlescent mistletoe, growing high up in the canopy, was a sure sign a tree had been specially selected by the gods.

Mistle thrush

Watching over the mistletoe, jealously guarding its opaline berries, is the mistle thrush. Like a songbird on steroids, this brown spotted bird is the largest of all the thrushes, and well known for its bullish defence of berry trees in freezing winter. Commonly called after its penchant for mistletoe, even its scientific Latin name – *Turdus viscivorus* – means mistletoe-devourer. From high in a berry tree, the mistle thrush defends its resources to the hilt. If other birds – from wood pigeons to waxwings – attempt to raid its winter larder, the mistle thrush flexes its muscles, driving away the competition with a chase and warning cry.

The mistle thrush's ballad is one of the few heard in a snow-covered landscape, a fruity melodious call not unlike the blackbird. It also keeps singing in rainy or cold weather, a doughty trait that once earned it the nickname of 'storm cock' or 'January joy'. If startled, however, or protecting its berries, the mistle thrush changes its tune to an aggressive chatter. It's an odd sound – a football rattle, perhaps, or a thumbnail across the teeth of a comb – but it seems to do the trick. Few birds that take on a mistle thrush make the same mistake twice and soon learn to steer clear of a defended tree. Its ancient name in Anglo-Saxon was the *scríc* or screech; in some parts of the country the mistle thrush is still known as the shrite, screech thrush or screaming mavis.

Despite its name, the mistle thrush isn't only fond of mistletoe. It happily demolishes seeds, fallen fruit, worms, insects and berries of all kinds, including rowan, ivy, yew, hawthorn and holly. For Victorian ornithologists the mistle thrush was also called the holm thrush, a centuries-old word for holly, especially in northern English counties and Scotland, where mistletoe rarely grows. This richly nutritious diet keeps mistle thrushes in peak condition – most individuals grow to around the same length as a long school ruler, with an even wider wingspan.

A bird that prefers free-standing trees in an open landscape, it is most commonly seen in parkland, orchards, farmland, mature gardens and woodland pasture. If a mistle thrush finds a particularly fruitful tree, it'll guard it year after year between autumn and spring, only to stop when the urge to breed becomes too distracting. Beady-eyed birdwatchers have also noticed a sneaky behaviour employed by the mistle thrush: although it will rigorously defend its berry-laden tree over winter, it'll conserve its food supply by dining elsewhere, usually on windfall fruit and undefended berry trees. By keeping its own berry tree full of fruit, and eating out instead, the mistle thrush can eke out its winter food supply, often making it last well into next spring. Truly a case of 'what's yours is mine and what's mine is my own' if there ever was one.

Robins

For all its Christmas card appeal, the robin is a bit of a bruiser.
A fiercely territorial bird, its red breast is a warning sign to other
robins to keep well away. If challenged, it pipes up an aggressive
tune and puffs out its chest, often swaying its body from side to side.
If all that posturing fails to intimidate, it raises the stakes and attacks.
Robin battles can be alarming tussles to witness – beaks locked,
feathers flying, tumbling and flapping, sometimes to the death. As
many as one in ten of all robins perish as a result of their skirmishes.
And, like any venerable fighter, the winner wears its victories on
its chest. The older and more experienced the robin, the larger the
surface area of their red breast – a signal to younger, less battle-
hardened robins to keep a safe distance.

Garden birds

Winter can be a lean time for wild birds, and yet there's plenty of debate about whether feeding garden birds is a good idea. While many avian species undoubtedly benefit from the nuts, seeds and fat we leave out, it's not clear whether the effect is positive across the board. Some conservationists worry that the types of bird that flourish with garden feeding outcompete other less gregarious species. Or, that piles of food on bird tables and feeders can inadvertently spread disease and change the behaviour and delicate balance of species in an environment.

So what's a bird lover to do? The key seems to be variety in both menu and delivery. Different species prefer different foods of different sizes. Finches and siskins, for example, love tiny nyjer seeds, while black sunflower seeds are irresistible to chaffinches and tits. Robins and blackbirds often favour mealworms, while thrushes, redwings and fieldfares devour windfall apples.

How you put out food also makes a difference. Large numbers of birds that congregate in the same location, at the same few feeders, are more likely to spread disease; especially if they come into regular contact with other bird species they wouldn't ordinarily meet. By offering different food types at separate places around the garden, in species-specific feeders, you can reduce the likelihood of birds with different diets having to jostle for space. Feeders should also be filled little and often, with just enough food to last between one and two days.

Best of all, however, is to create a garden that helps sustain wild bird populations naturally. By growing plants that provide berries, nuts, hips, seeds and habitats for insects, you can provide a succession of meals right through until spring. Plants with seedheads, such as sunflowers and teasels, are ideal, as are berry-laden plants such as hawthorn, ivy, cotoneaster and rowan. Plants that produce fruits and hips over late autumn and winter – such as crab apples, dog roses and rugosa roses – provide bountiful buffets of both flesh and seeds. Even the Christmas favourites, holly and ivy, are a gift, offering rich rewards right though the season.

Moon gazing

For thousands of years, our ancestors used the moon to mark time. Indeed, its very name speaks of its primary purpose as a lunar calendar. The word for 'moon' is strikingly similar in many different modern and historic languages, from the Greek *mene* to the Latin *mensis*, Viking *mani* or Persian *mah*. This likeness probably originated from an early language, shared across Central Asia and Europe, around 6,000 years ago. Scholars call it 'Proto-Indo-European' or PIE, a long-winded title for what was essentially an ancient mother tongue and the ancestor of many different languages today. And, in PIE, the word that inspired all our modern names for moon was *me*, which meant 'to measure'. The moon, since the dawn of humanity, has always measured time.

But how, exactly? If you stared up at the moon every evening for a month, you'd see it had a slightly different shape. From a full circle to a thin sliver of a crescent, what we see changes on a nightly basis. These ever-shifting shapes are called 'phases' and are created by the moon's orbit. As it journeys around the Earth, the moon reflects different amounts of light from the sun depending where it is in the cycle.

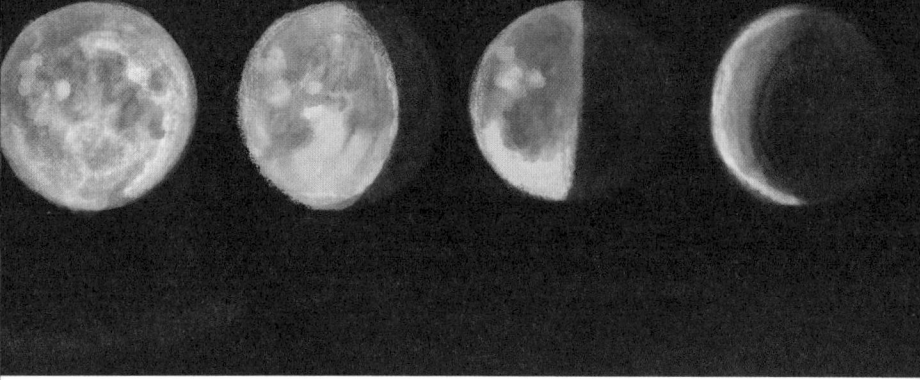

It takes 29.5 days for the moon to take a full trip around the Earth. The beginning of the month is marked by a 'new moon', when the moon sits directly between Earth and the sun. At this point, we see no reflected light and this makes the moon appear completely black. As the month progresses, we start to see more of the moon. At first, we catch a glimpse of a banana-shaped 'waxing crescent'. This is followed by a 'first quarter' and then an almost circular 'waxing gibbous'. When we hit the halfway mark of the month, we finally see a glorious, perfectly round full moon.

The second half of the month sees the moon's light disappearing or 'waning' back to complete darkness. Directly after the full moon we have a 'waning gibbous', then a 'last quarter' and finally a narrow 'waning crescent'. In the northern hemisphere, these phases slowly illuminate the moon from right to left. In the southern hemisphere, the direction is reversed.

Sometimes a full moon can seem much bigger than normal. This is called a 'super moon' and there are usually about four of these a year. As the moon orbits the Earth it doesn't travel in a perfect circle – its journey is elliptical. As it whizzes around the Earth, the distance between the moon and us can vary from between 400,000km and 360,000km (250,000 miles and 225,000 miles). The closer the moon is to Earth, the larger it appears.

Although it may sound counterintuitive, a full moon isn't the optimum time for moon gazing. This is because the reflected light is too bright and almost all the moon's stark features are obscured. Instead head out during a wintry crescent, quarter or gibbous phase and focus on the moon's 'terminator' or twilight zone, the line between light and shadow. There, the contrast between the brightness and darkness will reveal the moon's craters and crags in all their primeval glory.

Barn owl

Despite its snowy white connotations, the barn owl is not particularly suited to winter. For a bird so closely associated with the countryside, it is also deeply cosmopolitan. The most widespread wild land bird species in the world, the barn owl is probably more at home in the balmy Indian subcontinent or Pacific Islands than in the hayloft of a farmer's chilly barn. Indeed, more species of barn owl live in the tropics and subtropics than in any temperate region.

Barn owls, for all their feathered magnificence, are poorly insulated and quickly lose body heat. Snow on the ground might look picturesque but it makes hunting for mice and other small mammals doubly hard work. The period between November and early spring can be an obstacle course for the hungry barn owl. Flying on a frozen night, looking for prey, uses up valuable energy. And it's not guaranteed that every foray will be a success. Numbers of small mammals also nosedive during the winter, leaving skinny pickings for a bird in need of sustenance.

But what is traditionally a challenging time of year for the barn owl is one of the best for spotting them, or 'owling'. Barn owls are notoriously elusive and only come out during the night or in the crepuscular twilight hours. But during winter and early spring they are often forced to extend their hunting trips into the bright daylight. Those precious few extra hours can mean the difference between life and death and, for the keen owl twitcher, present one of the few opportunities to see them up close. Barn owls are also creatures of habit. They often hunt over the same few fields or woodland edges, visiting each place at roughly the same time of day. You may even see a barn owl on enough occasions to start to piece together its routine and favourite resting places.

As their name suggests, barn owls are at home in the farmed landscape. The agricultural way of life, with its open fields, verges and hedgerows, suits its style of hunting. Using a technique known as 'quartering', the barn owl glides silently over farmland just a few metres off the ground, flying back and forth to ensure no patch of land is left unsearched. If a small creature makes even the most infinitesimal sound, the barn owl will hear it and lock on to its target. For a brief moment it pauses, mid-air, before plunging to the ground headfirst only to swing its legs forward, talons outstretched, at the very last second. In winter, when the barn owl is only feeding itself and not a small family of owlets, you may also see a hunting behaviour called 'mantling'. Once the barn owl lands on its prey, like a vampire about to ravish its victim, it pulls its wings around its kill. Secretive and guarded, the barn owl can then enjoy its much-needed winter feast in peace.

Foxes mating

In the dead of a winter's night, you can hear screams that send a chill to your core. January and February are the mating season of the red fox, a time when both urban streets and village greens are kept awake by blood-curdling shrieks. In midwinter the vixen is on heat for a few, urgent days. It's her frightful cries you hear, piercing the darkness – but don't be alarmed. These drawn-out howlings are, in fact, calls of passion to the mate of her choice. It used to be thought that these shrieks were hollers of agony. When foxes mate, they become locked in a position called a 'copulatory tie', both unable to extricate themselves for up to an hour. Although painful to look at, it's a cunning strategy to ensure the male's sperm is safely implanted. And, when it's all over, he'll continue his jealous guarding, keeping her close and making sure no one else catches her eye.

Badgers

One animal that excels at torpor is the badger. When the winter weather takes a turn for the worse, she shuffles down into her sett and falls into a deep, relaxing slumber. Her loose-fitting pelt, like baggy striped pyjamas, is the perfect nightwear. Throughout autumn, the badger has been eating furiously and built up a thick layer of fat under her slackly anchored skin. She's particularly fond of earthworms and has eaten thousands over the past three months to boost her fat reserves up to nearly a third of her total body weight. She'll need it too – this cozy layer of insulation will help buffer any periods of food scarcity and keep her warm should the temperature really tumble.

At the beginning of winter, the badger is at her heaviest. As the season progresses, she spends less time eating and more time snoozing underground, slowly using up those essential calories. Winter is also the time of year when she's most likely to give birth. Badgers can mate throughout the year but have a canny reproductive trick, called delayed implantation, which means the female gives birth to only one litter a year, usually in February.

Her tiny litter of cubs – usually just two or three – is born underground and stays there, in warm darkness, until spring. With such a comfortable subterranean nursing chamber, filled with plump bedding material, it's no wonder the cubs are reluctant to venture above ground. At eight weeks old, they might stand in the entrance to their sett and nervously survey their surroundings. Only four weeks after that, however, and weaning begins in earnest and soon they're foraging for themselves. By the beginning of the following December, the badger cubs are almost full grown. Now, wearing their own pair of baggy, loose-fitting pyjamas, and with a tummy full of berries and invertebrates, they too will spend the winter dreaming of earthworms somewhere deep, deep under the snow.

Birds of prey

In places that have relatively mild winters, such as Britain, many birds are permanent residents. Rather than fly to warmer climes to wait out the frigid months, plenty of species stay put and adjust their diets to a new, seasonal menu. Birds of prey, which need plenty of protein-rich food to survive, can be surprisingly broadminded about what they'll hunt when the usual fare has died off or gone into hibernation. Buzzards, for example, are famously unfussy when it comes to winter grub. For such large birds, they can make a meal of the tiniest morsel. In summer, the buzzard has access to plenty of rodents, amphibians and rabbits. In winter, when many small mammals and frogs hunker down out of sight, or hide under the cover of snow, it boosts its diet with earthworms, beetles and other insects. It'll also take opportunistic meals of crows, pigeons and carrion. Buzzards have even been known to keep watch on molehills, ready to pounce should an unsuspecting mole break cover.

It's a similar story for the common kestrel. During the warmer months, the kestrel's favourite prey is the vole. But if winter numbers of voles are down, a kestrel might switch to earthworms, small birds and, in some warmer countries, even bats. For one bird, however, the winter presents few challenges. The peregrine falcon is the fastest animal on Earth, and a sublime hunter whatever the season. It also travels to where winter birds flock. During this time of year, it often heads to estuaries and coastlines to find crowds of ducks, gulls and waders. Murmurations of starlings also prove too tempting to ignore, as are flocks of fieldfares and wood pigeons. In fact cities, with their constant supply of feral pigeons, starlings, thrushes and sparrows, are some of the best places to see peregrine falcons. London alone is thought to have at least 30 breeding pairs. Whether it's a church spire or city landmark, many peregrine falcons across mainland Europe and Britain have swapped cliffs and rocky ledges for urban plenty and high-rise living.

Catkin

Catkins start to appear in the depths of winter, when many trees are otherwise naked. Not only are they a welcome harbinger of spring but their true purpose is reproduction, wafting great clouds of pollen into the wind when a tree's leaves can't get in the way. While the hazel often bursts into life in January, with its dangling golden tails, the alder isn't far behind. But it can be tricky to tell the two apart. Look for the scorched appearance of alder catkins, like a hazel catkin held too close to the fire. Many trees have catkins, including the silver birch, aspen, poplar and pedunculate oak and the name, rather adorably, means 'kitten'. For the cuddliest catkin of all, however, you'll have to wait until early spring. The silver fur catkins of the goat and grey willows are just too strokable to ignore. No wonder these have earned themselves the special feline nickname 'pussywillow'.

Rooks

As winter draws to a close, high up in the treetops all hell is breaking loose. Rooks clamour to build nests and these garrulous, sociable birds can be heard from afar, yelling loudly to each other across the branches like workers on a construction site. A pair of rooks will team up and often find the remnants of a doer-upper, one of last year's nests that needs renovating. While the male flies back and forth, dragging materials in his beak, the female is the true project manager, carefully placing each twig, leaf and piece of straw to create the perfect home for the next months' batch of eggs. And, like all resourceful renovators, rooks don't always need fresh building supplies to finish the job. 'Borrowing' from each other's nests is common practice, pilfering twigs and soft lining material from fellow rooks to complete their own projects on time. Maybe that's what all the shouting is about.

Early flowers

For all its beauty, winter can sometimes feel interminable. After the colour and mayhem of Christmas have faded, the gap between December and the arrival of spring can seem like a chasm. And yet, thanks to a handful of early flowering plants that push through the frozen ground at this time of year, we're soon reminded that life is never far from the surface. Just a smattering of tiny flowers on a gloomy winter's day can lift the spirits as much as a gardenful of gaudy blooms only a few months later.

One of the stalwarts of the season is the hellebore. It flourishes in the depths of midwinter, when almost all other life in the garden has withered, earning it the affectionate, seasonal name of the Christmas Rose. So too the snowdrop, one of the earliest flowers to appear in the New Year. In mild winters, it can peek through the soil as early as January, often under the protection of a thick woodland canopy. But most bloom in February, an event celebrated by its old folk name 'Fair Maids of February'. The faithful also called them 'Candlemas Bells', a pretty moniker derived from one of the oldest Christian festivities, February's Candlemas or Feast of the Presentation. Not long after, crocuses begin to embroider the ground. When sunshine is at a premium, the warm eggy yellows of these cheerful flowers couldn't be more welcome. And, as if nature had a colour wheel at her disposal, more crocuses appear in complementary deep violets and pure, pure whites.

But for all their welcome colour these three winter flowers also garnered superstitions. Flowers and plants that appeared at unusual times of the year, defying the natural order of things, often invited curious folk beliefs. Crocuses, for example, were believed to sap your strength as you plucked them – only those people in truly rude health should gather these seemingly innocuous winter flowers. Snowdrops were also considered unlucky, but only if you brought them into the house. Indeed, such was the fear of an indoor bouquet of these winter blooms, that many believed they would cause a death in the family before the year was out. It didn't help the poor snowdrop that some thought it resembled a tiny corpse in a white shroud, an imaginative yet gloomy interpretation for such a sweetly perfect flower.

Most fearfully respected of all, however, was the hellebore and not without good reason. Although it bloomed over winter, its real potency came from its toxicity. Ancient medicinal texts often suggested hellebore as a treatment for insanity, melancholy, and epilepsy. Others thought it might sharpen the wits and strengthen the brain. It is, in fact, rather poisonous and not a plant to be fooled with; if ingested, it acts as a violent and crippling purgative. One of the most bizarre beliefs about the hellebore, however, was that once its roots were dried and powdered, they could make you invisible. A quick toss of hellebore dust into the air and – whoosh – you disappeared into the thin air.

Animal tracks

Winter's mammals can seem to disappear like magic. As the snow falls, dampening the sounds and sights of the landscape, it can feel as if many creatures have vanished. And yet, for all their shy reluctance, our native wildlife often leaves tell-tale traces in the mud or snow cover. Learn to identify these unmistakable footprints and you'll never look at the ground in the same way again.

Red Fox

The red fox's pawprints appear deceptively similar to a dog's but the carpal footpad is a dead giveaway. Dogs have very large carpal footpads compared with the soft parts under their toes (called digital pads). On a fox, however, the carpal and toe pads are the same size. The fox's foot also leaves an oval footprint, compared to the broader, circular paws of the domestic dog, and the fox's claw marks are often visible. Look for tracks approximately 3.5cm (1.4in) wide and 5cm long (2in).

Deer

Often said to look like a broken heart, the footprint of the deer consists of two pointed oblongs. Deer have cloven hooves – a foot divided into two parts called cleaves – that leave a long, sharp-edged print with a narrow gap in between. Sizes can vary from the tiny muntjac's 3cm (1.2in) hoof to the impressive 9cm-long (3.5in) cleaves of a red deer stag.

Wild Boar

Wild boar tracks are often mistaken for those of sheep or goats, but there are two clues that can help. One is location – wild boar tracks tend to be found in woodland – but the other, more definitive difference is the marks made by a swine's dew claws. At the back of each footprint you'll see a pair of smaller marks, pointing outwards. These are made by the wild boar's dew claws that, unlike those of other species, make contact with the ground as they walk. Larger wild boar footprints can reach 8.5cm (3.3in) long and 7cm (2.75in) wide.

Brown Hare

These graceful, introverted creatures leave very distinctive footprints. While the hare's front paws are small, and placed daintily one in front of the other, its back feet are huge and planted almost parallel as they run. At speed, a hare strikes down its front feet on the ground first and then brings its hind legs even further in front, creating its characteristic back-feet-first footprint. Search for front pawprints 2.5cm (1in) wide and 3.5cm (1.4in) in length.

Squirrel

While squirrels can scamper around casually on all fours, they often leave their most distinctive prints when they bound. Just like the hare, the squirrel places its small front paws on the ground first, and then brings forward its larger hind feet, landing so they straddle the forefeet or are slightly in front. Squirrels have four toes on their front feet and five on their hind feet, leaving prints that can look like miniature hands pressed into the snow.

Directory

Birds & Bird Song

RSPB
www.rspb.org

British Trust For Ornithology
www.bto.org

XENO-CANTO
(Bird Song Identification)
www.xeno-canto.org

Swift Conservation
www.swift-conservation.org

The Barn Owl Trust
www.barnowltrust.org.uk

The World Owl Trust
www.owls.org

Hawk Conservancy Trust
www.hawk-conservancy.org

Starlings in the UK
www.starlingsintheuk.co.uk

Wildfowl & Wetlands Trust
www.wwt.org.uk

Birdlife International
www.birdlife.org

Mammals

Mammal Society
www.mammal.org.uk

The Hare Preservation Trust
www.hare-preservation-trust.com

The Wildlife Trusts
www.wildlifetrusts.org

People's Trust for
Endangered Species
www.ptes.org

Bat Conservation Trust
www.bats.org.uk

The British Deer Society
www.bds.org.uk

Red Squirrel Survival Trust
www.rsst.org.uk

The British Hedgehog
Preservation Society
www.britishhedgehogs.org.uk

Hedgehog Street
www.hedgehogstreet.org

Badger Trust
www.badgertrust.org.uk

Invertebrates

Bumblebee Conservation Trust
www.bumblebeeconservation.org

Buglife
www.buglife.org.uk

Royal Entomological Society
www.royensoc.co.uk

Pollinator Partnership
www.pollinator.org

UK Moths
www.ukmoths.org.uk

Butterfly Conservation
www.butterfly-conservation.org

The Earthworm Society
of Britain
www.earthwormsoc.org.uk

Amphibians & Reptiles

Froglife
www.froglife.org

Amphibian and Reptile
Conservation
www.arc-trust.org

Societas Europaea
Herpetologica
www.seh-herpetology.org

Center For Snake Conservation
www.snakeconservation.org

Skies & Weather

Met Office
www.metoffice.gov.uk

Royal Meteorological Society
www.rmets.org

The European Centre for
Medium-Range Weather
Forecasts
www.ecmwf.int

Snow Crystals
(Snow Science)
www.snowcrystals.com

Dark Skies & Stars

Dark Skies Festival
*www.darkskiesnationalparks
.org.uk*

Dark Sky Discovery
www.darkskydiscovery.org.uk

Royal Astronomical Society
www.ras.org.uk

Google Sky
www.google.com/sky

Woodland & Hedgerows

Woodland Trust
www.woodlandtrust.org.uk

The Tree Register
www.treeregister.org

The Arboricultural Association
www.trees.org.uk

The Conservation Volunteers
www.tcv.org.uk

Reforest Britain
www.reforestbritain.com

Trees for Cities
www.treesforcities.org

Hedgelink
www.hedgelink.org.uk

The Fungus Conservation Trust
www.fungustrust.org.uk

Fungi Foundation
www.ffungi.org

British Pteridological Society
(Ferns)
www.ebps.org.uk

European Networks for Private
Land Conservation
www.enplc.eu

Ponds & Waterways

The Rivers Trust
www.theriverstrust.org

Canal & River Trust
www.canalrivertrust.org.uk

Freshwater Habitats Trust
www.freshwaterhabitats.org.uk

Uk Centre for Ecology
and Hydrology
www.ceh.ac.uk

Wildflowers & Verges

Plantlife
www.plantlife.org.uk

The Wildlife Trusts
www.wildlifetrusts.org

RHS
www.rhs.org.uk

Eden Project
www.edenproject.com

Save Our Magnificent Meadows
www.magnificentmeadows.org.uk

Citizen Science

Nature's Calendar
naturescalendar.woodlandtrust.org.uk

eBird
www.ebird.org

Mammal Web
www.mammalweb.org

iNaturalist
www.inaturalist.org

UK Beetle Recording
www.coleoptera.org.uk

Spring Alive
www.springalive.net

Moth Night
www.mothnight.info

National Moth Recording Scheme
www.mothrecording.org

Big Butterfly Count
bigbutterflycount.butterfly-conservation.org

Toad Patrol
www.froglife.org

Natural History Museum
www.nhm.ac.uk

General

National Trust
www.nationaltrust.org.uk

Rewilding Britain
www.rewildingbritain.org.uk

CPRE, The Countryside Charity
www.cpre.org.uk

National Biodiversity Network
www.nbn.org.uk

International Union for Conservation of Nature
www.iucn.org

Earthwatch Europe
www.earthwatch.org.uk

Restore Our Planet
www.restoreourplanet.org

National Parks UK
www.nationalparks.uk

Soil Association
www.soilassociation.org

BBC Earth
www.bbcearth.com

BBC Countryfile
www.countryfile.com

Index

About the author

Sally Coulthard has spent the past two decades writing about her favourite things – nature, history and craft. Many of her books delve into the traditions of rural life and the natural world, the people, plants and creatures who make the countryside tick. Sally's work often weaves together different disciplines, pulling threads from nature writing, social history, archaeology and folklore to bring her diverse subjects to life. She also runs a smallholding deep in the North Yorkshire countryside, where a gaggle of unruly animals runs rings around her.

Acknowledgements

Harriet and Emily – what a joy to work with such fiercely bright and furiously funny people. You are the best. It is *always* a pleasure to make lovely books with you. And unbridled thanks to Gemma – illustrators have such a tough time interpreting the whims of an author but you well and truly nailed it.

Managing Director Sarah Lavelle
Senior Commissioning Editor Harriet Butt
Editorial Assistant Ellie Spence
Senior Designer Emily Lapworth
Illustrator Gemma Koomen
Head of Production Stephen Lang
Production Controller Sabeena Atchia

Published in 2024 by Quadrille,
an imprint of Hardie Grant

Quadrille
52–54 Southwark Street
London SE1 1UN
quadrille.com

Cataloguing in Publication Data: a catalogue record
for this book is available from the British Library.

Text © Sally Coulthard 2024
IIllustrations © Gemma Koomen 2024
Design © Quadrille 2024

ISBN 978 1 83783 154 8

Printed in China and using soy inks.